Erwin Schrödinger's book *What is Life?* had a tremendous influence on the development of molecular biology, stimulating scientists such as Watson and Crick to explore the physical basis of life. Much of the appeal of Schrödinger's book lay in its approach to the central problems in biology – heredity and how organisms use energy to maintain order – from a physicist's perspective.

At Trinity College, Dublin a number of outstanding scientists from a range of disciplines gathered to celebrate the fiftieth anniversary of *What is Life?*, and following Schrödinger's example fifty years previously, presented their views on the current central problems in biology. The participants included Jared Diamond, Christian de Duve, Manfred Eigen, Stephen Jay Gould, John Maynard Smith, Roger Penrose and Lewis Wolpert.

What is Life? The Next Fifty Years

From left to right: Stephen Jay Gould, John Maynard Smith, Lewis Wolpert, Stuart Kauffman, Roger Penrose, Gerald Edelman, Walter Thirring and Leslie Orgel. (Photograph by Margaret Worrall.)

What is Life? The Next Fifty Years

Speculations on the future of biology

EDITED BY

MICHAEL P. MURPHY
Department of Biochemistry, University of Otago, Dunedin

LUKE A. J. O'NEILL
Department of Biochemistry, Trinity College, Dublin

of Cambridge
e CB2 1RP
40 West 20th Street, New York, NY 10011-4211, USA
10 Stamford Road, Oakleigh, Melbourne 3166, Australia

First published 1995

Printed in Great Britain at the University Press, Cambridge

A catalogue record for this book is available from the British Library

Library of Congress cataloguing in publication data

What is life? : the next fifty years : speculations on the future of biology / edited by Michael P. Murphy and Luke O'Neill.
p. cm.
Contains most of the contributions presented at a conference held at Trinity College, Dublin, from Sept. 20–22, 1993.
Includes bibliographical references.
ISBN 0 521 45509 X (hc)
1. Biology – Philosophy – Congresses. 2. Biology – Congresses. 3. Life (Biology) – Congresses. 4. Schrödinger, Erwin, 1887–1961. What is life? – Congresses. I. Murphy, Michael P. (Michael Patrick), 1963– . II. O'Neill, Luke.
QH331.W465 1995
574'.01 – dc20 94–49438 CIP

ISBN 0 521 45509 X hardback

RO

Contents

Contributors

Ruth Braunizer
A-6236 Alpbach 318, Tirol, Austria

Christian de Duve
ICP 75.50, Avenue Hippocrate 75, B-1200 Brussels, Belgium

Jared Diamond
Department of Physiology, UCLA Medical Center, 10833 Le Conte Avenue, Los Angeles, CA 90024-1751, USA

Manfred Eigen
Max Planck Institut für Biophysikalische Chemie, Postfach 2841, D-37077 Göttingen, Germany

Stephen Jay Gould
Museum of Comparative Zoology, Harvard University, 26 Oxford Street, Cambridge MA 02138, USA

Hermann Haken
Institute for Theoretical Physics & Synergetics, University of Stuttgart, Stuttgart, Germany

Stuart A. Kauffman
Santa Fe Institute, 1660 Old Pecos Trail, Suite A, Santa Fe, NM 87501, USA

James J. Kay
Environment and Resource Studies, University of Waterloo, Waterloo, Ontario, Canada N21 3G1

J. A. Scott Kelso
Program in Complex Systems & Brain Sciences, Center for Complex Systems, Florida Atlantic University, Boca Raton, FL, USA

John Maynard Smith
Biology Building, The University of Sussex, Falmer, Brighton, Sussex BN1 9QG, UK

Michael P. Murphy
Department of Biochemistry, University of Otago, Box 56, Dunedin, New Zealand

Luke A. J. O'Neill
Department of Biochemistry, Trinity College, Dublin 2, Ireland

Roger Penrose
Mathematical Institute, 24–29 St Giles, Oxford OX1 3LB, UK

Eric D. Schneider
Hawkwood Institute, P.O. Box 1017, Livingston, MT 59047, USA

Eörs Szathmáry
Department of Plant Taxonomy and Ecology, Eötvos University, Budapest, Hungary

Walter Thirring
Institut für Theoretische Physik, Universität Wien, Boltzmanngasse 5, A-1090 Wien, Austria

Lewis Wolpert
Department of Anatomy and Developmental Biology, University College and Middlesex School of Medicine, Windeyer Building, Cleveland Street, London W1P 6DB, UK

Preface

A conference was held at Trinity College, Dublin from 20 to 22 September 1993 to celebrate the fiftieth anniversary of Erwin Schrödinger's lectures *What is Life?* At this conference scientists from a number of disciplines speculated, in the manner of Schrödinger's original lectures, on the development of biology over the next fifty years. This volume contains most of these contributions. In addition, there are a few contributions from scientists who were unable to speak at the conference.

The editors thank Otago University; the Wellcome Trust; the Austrian Embassy, Dublin; the Biochemical Society, London; TCD Association and Trust; the Dublin Institute for Advanced Studies; the Royal Irish Academy; BioResearch Ireland; the British Council; Biotrin International; and Pharmacia Biotech for their generous support. It is a pleasure to acknowledge the help and advice we received throughout from Dr Joe Carroll, Dean of Science, Trinity College, Dublin; Dr Margaret Worrall, Newman Fellow, University College, Dublin; Dr Tim Mantle, Department of Biochemistry, Trinity College, Dublin; Ms Alex Anderson, Trinity College, Dublin; Professor John Lewis, Dublin Institute for Advanced Studies; Professor David McConnell, Department of Genetics, Trinity College, Dublin; Professor Keith Tipton, Department of Biochemistry, Trinity College, Dublin; Associate Professor Merv Smith, Department of Biochemistry, University of Otago, Dunedin; Dr Garret Fitzgerald, Dublin; and Mr Louis le Brocquy, Carros, France.

1

What is Life? *The next fifty years. An introduction*

MICHAEL P. MURPHY[1] and LUKE A. J. O'NEILL[2]

[1]*Department of Biochemistry, University of Otago, Dunedin*

[2]*Department of Biochemistry, Trinity College, Dublin*

This book is the result of a conference held in Trinity College, Dublin in September 1993 which commemorated the fiftieth anniversary of a series of lectures entitled *What is Life?*, given in Trinity College in 1943 by Erwin Schrödinger. Schrödinger, a Nobel-prize-winning physicist and one of the founders of quantum mechanics, had come to Dublin in 1939 at the invitation of Éamonn de Valera, the Taoiseach (Prime Minister) of Ireland to take up a Chair of Theoretical Physics at the newly founded Dublin Institute for Advanced Studies (Moore, 1989; Kilmister, 1987). The invitation followed his dismissal from the Chair of Theoretical Physics at the University of Graz after the *Anschluss*. Dublin suited Schrödinger and he fitted in well, becoming a leading personality in the intellectual life of the city. He remained in Dublin until his return to Austria in 1956, where he died five years later.

Schrödinger had broad intellectual interests and while in Dublin he explored areas of philosophy and biology as well as continuing to work in theoretical physics. In this volume we are concerned with Schrödinger's thinking on biology. In *What is Life?* Schrödinger focused on two themes in biology: the nature of heredity and the thermodynamics of living systems. Delbrück was an influence on Schrödinger's views on heredity while Boltzmann stimulated much of his work on the thermodynamics of living systems. For the first presentation of his thinking on biology Schrödinger chose a public lecture. An annual public lecture is a statutory obligation of the Dublin Institute for Advanced Studies and in February 1943 Schrödinger gave a series of three lectures to a general audience at Trinity College, Dublin. These lectures were popular with Dubliners and over four hundred stayed through the entire series. No doubt part of their popularity was the provocative title and the restricted entertainment available during the 'emergency', as the Second World War was called in neutral Ireland, but in addition Schrödinger was a gifted public speaker who could captivate an audience.

After their publication by Cambridge University Press (Schrödinger, 1944) these lectures had considerable impact internationally. The book was widely read and became one of the most influential 'little books' in the history of science (Kilmister, 1987). Surprisingly, in spite of the widely acknowledged influence of this book on the founders of molecular biology (Judson, 1979), the precise role played by *What is Life?* is still disputed (Judson, 1979; Pauling, 1987; Perutz, 1987; Moore, 1989). Undoubtedly, part of the appeal and influence of the book was its clear prose and the persuasiveness of the arguments. Schrödinger, portraying himself as a 'naive physicist', made it clear how living systems could be thought of in the same way as physical systems. Clearly this approach was already widespread, but *What is Life?* popularized it and encouraged physical scientists that the time was ripe to consider biological problems.

What about the actual ideas expressed in the book? Schrödinger discussed two themes based on his thinking on heredity and thermodynamics. In one of these themes, usually termed the 'order from order' theme, Schrödinger discussed how organisms pass on information from one generation to the next. As a basis for this discussion about the gene he used the well-known paper by Timoféeff-Ressovsky, Zimmer and Delbrück (1935) on mutation damage to fruit flies from which the size of the gene was calculated to be about 1000 atoms. The problem faced by the cell was how a gene this size could survive thermal disruption and still pass on information to future generations. Schrödinger proposed that to avoid this problem the gene was most probably some kind of aperiodic crystal which stored information as a codescript in its structure. As is well known, this prophetic statement has been shown to be true by work on the structure of DNA which led to the central dogma of molecular biology. The second theme covered by Schrödinger was 'order from disorder'. The problem faced by organisms was how to retain their highly improbable ordered structure in the face of the second law of thermodynamics. Schrödinger pointed out that organisms retain order within themselves by creating disorder in their environment. However the term 'negentropy', which he coined for this process, has not been well received by other scientists (e.g. Pauling, 1987).

In the fifty years since Schrödinger's lectures we have become accustomed to the 'order from order' theme and much of the astonishing success of molecular biology over the past fifty years can be seen as working out the implications of this idea. It is on this that much of the reputation of *What is Life?* is based. The 'order from disorder' theme has generally been considered of less significance. However, now that work on the thermodynamics of systems removed from equilibrium and dissipative structures is being applied to living systems the importance of this theme may reassert itself. Perhaps fifty years from now *What is Life?* will be seen as prophetic for its treatment of the thermodynamics of living systems rather than for the prediction of the structure of the gene.

While the influence of *What is Life?* is acknowledged, the ideas expressed have been criticized as unoriginal or wrong (Pauling, 1987; Perutz, 1987) by some while defended by others (Moore, 1987; Schneider, 1987). It is true that much that was explicit in *What is Life?* was implicit in earlier work. However, these criticisms perhaps miss a major aspect of the uniqueness of *What is Life?*: that a physicist straying from his area of expertise into a field not his own could stimulate research. This interdisciplinary posing of provocative questions is not usual in science and in *What is Life?* the musings of a physicist have acted as an inspiration to subsequent researchers. It is in this spirit that we commemorate the lectures fifty years ago of Erwin Schrödinger. In doing this we have gathered together a number of articles in which scientists speculate on the future of biology. Much expressed in this volume may turn out to be wrong; however, we believe that this exploratory spirit is the best way to commemorate the publication fifty years ago of *What is Life?*

REFERENCES

Judson, H. F. (1979). *The Eighth Day of Creation: Makers of the Revolution in Biology*. New York: Simon & Schuster.

Kilmister, C. W. ed. (1987). *Schrödinger: Centenary Celebration of a Polymath*. Cambridge: Cambridge University Press.

Moore, W. J. (1987). Schrödinger's entropy and living organisms. *Nature* **327**, 561.

Moore, W. J. (1989). *Schrödinger: Life and Thought*. Cambridge: Cambridge University Press.

Pauling, L. (1987). Schrödinger's contribution to chemistry and biology. In *Schrödinger: Centenary Celebration of a Polymath*, ed. C. W. Kilmister, pp. 225–233. Cambridge: Cambridge University Press.

Perutz, M. F. (1987). Erwin Schrödinger's What is Life and molecular biology. In *Schrödinger: Centenary Celebration of a Polymath*, ed. C. W. Kilmister, pp. 234–251. Cambridge: Cambridge University Press.
Schneider, E. D. (1987). Schrödinger's grand theme shortchanged. *Nature* **328**, 300.
Schrödinger, E. (1944). *What is Life? The Physical Aspect of the Living Cell.* Cambridge: Cambridge University Press.
Timoféeff-Ressovsky, N. W., Zimmer, K. G. & Delbrück, M. (1935). *Nachrichten aus der Biologie der Gesellschaft der Wissenschaften Göttingen* **1**, 189–245.

2

What will endure of 20th century biology?

MANFRED EIGEN
Max Planck Institut für Biophysikalische Chemie, Göttingen

Acknowledgements

The original version of this lecture was published in 1993 in the book *Man and Technology in the Future*, a summary of an international seminar arranged by the Royal Swedish Academy of Engineering Sciences, Stockholm, Sweden.

'QUO VADIS HUMANITAS?'

We find ourselves in the last decade of this century; no previous century has had such a profound effect on human life. Perhaps no century has produced such a level of apprehension and fear, anchoring them in the consciousness of man. One has become mistrustful. When a discovery becomes known nowadays, the first question is not, 'Of what use will it be to mankind?' (as in earlier times) but, 'What damage will it cause, and how will it diminish our well-being and health?' Our present state of well-being is bestowed upon us mainly owing to scientific knowledge; this has brought life expectancy up to 75 years, approaching the biologically natural age limit. At the beginning of this century, life expectancy was a mere 50 years and at the beginning of the previous century it was only about 40 years. In developing countries, the curve of life expectancy is also rising, although it lags about 50 years behind ours; meanwhile, our life expectancy is approaching an upper limit. Yet, as never before, we peer apprehensively into the future. This is despite the fact that in the political sector, some of the gravest and most grotesque developments instigated by humanity in this century appear to be in the process of rectification. It is unlikely to be decided in this last decade whether these changes are really for the better.

This decade not only brings the century to a close; it ushers in a new millennium. We feel impelled to reflect on the way we have come and on the road ahead. Our predicament becomes conscious in the question: 'Will humanity even survive to the end of the coming millennium?' Of the thirty or so generations that span a thousand years, we already have direct experience of two or three. These thirty generations may be listed with space to spare on a printed page; but, nonetheless, a thousand years defies our comprehension. What indeed could Charlemagne have predicted about our times? Proper experience of the past is essential for any extrapolation to the future but, even then, what is really new remains a surprise. In basic research, the situation is no different. New insights can open up whole continents of new opportunities. Moreover, all the things that shape our daily life depend essentially on discoveries and insights from the most recent past. All that we can really say about the future is almost a truism: changes in our way of life will be yet more radical in the coming millennium than they have been in that which is drawing to a close.

The world population is currently growing hyperbolically. How does hyperbolic differ from the exponential growth that is usually referred to in publications on this subject? Well, the latter involves successive doublings at equal intervals of time; with hyperbolic growth, these intervals become steadily shorter. A constant percentage rate of birth already yields exponential population growth, but, over and above this, an increasing percentage of people reach sexual maturity as a result of improved hygiene and medical care of infants and children in developing countries. The most recent doubling of the world population took only 27 years. There are now 5.5 billion of us on earth. If things continue according to the hyperbolic law, which has accurately described the increase of the past 100 years, there will be 12 billion people in 2020 and in 2040 the growth curve will tend asymptotically to infinity! I can see myself being quoted in the media: 'Scientist prophesies growth catastrophe in the year 2040.' Steady on now: the only prediction that I can make with certainty is that this will not take place; it cannot, since the resources of the earth are limited. We do not know where the coming century will lead us. Nevertheless, the really uncanny aspect of our predicament is not this fatalistic nescience. Much more disconcerting is the fact that we cannot derive anything from the present growth behaviour, not even in principle. Near such a singularity, even the smallest fluctuations can be amplified and come to have an enormous effect. Catastrophes, on a small scale or even of a global character, will limit the growth of world population. Such catastrophes are certainly not new to us. We know too that we stand helpless before them in their path. There is something amiss with our ethics, which is still matched to an epoch where human survival (or that of smaller demographic units) had to be secured through numerous offspring.

You may wish to interject that the population of industrial nations long ago reached equilibrium. In some countries it is even declining. Nonetheless, our population density is so great that, if it were to spread to the entire land mass, there would be a population of 30 to 40 billion people. According to a study by Roger Revell, that would be about the maximum number that could be maintained by mobilizing all conceivable planetary resources. An increase in food harvests over the entire earth to the local maximum when he wrote (corresponding to the corn harvest of the state of Iowa in the USA for instance) would be necessary just to barely feed such a population. There could be no prospect of general prosperity. The number calculated by Revell allows perhaps for a few regions of ample production, but in most regions there would be a catastrophic deficit. In this analysis, I have not even mentioned the environmental problems that are already getting out of control. Neither has mention been made of bottlenecks in the exploitation of resources and in energy production, nor of sanitation or medical emergencies.

This must suffice for an introduction. I wanted to describe the backdrop before which humanity's development will be played out. We should not lose sight of it when considering the future of science and our associated expectations, fears and hopes.

Turning now to the main topic, I will begin my exposition by taking stock of the current situation.

THE BIOLOGY OF THE 20TH CENTURY

One is indeed justified in proclaiming the second half of this century as the era of molecular biology, analogously to the first half as the age of atomic physics. In fact it was physicists who first took up the analysis of the concept of life, even if this initially led in the wrong direction. Pascual Jordan's *Physics and the Secret of Organic Life* from the year 1945 and most notably Erwin Schrödinger's 1944 book *What is Life?*, the event we are celebrating in this volume, are characteristic examples. Schrödinger's text was epoch-making, not because it offered a useful approach to an understanding of the phenomenon of life, but because it inspired new directions of thought.

Much of Schrödinger's prophetic content had long since been resolved by biochemists, but no one had previously so openly delved for basic principles. Nonetheless, it was not pure theoreticians who initiated the turn of the tide in biology and established the new science of molecular biology. They stood helpless in the face of the complexity of living things. Rather it was physicists who began to experiment in a radically new way, using our basic knowledge of the chemical nature of life processes as a springboard. There was Max Delbrück, a theoretical physicist of the Göttingen school who, inspired by Niels Bohr's complementarity principle, decided to investigate the molecular details of inheritance. This was the foundation of phage genetics. And then there was Linus Pauling, a physicist of Sommerfeld's school who sought a deeper understanding of the nature of proteins, the molecular executive of living cells. He discovered in the process essential structural elements, forming figuratively a seam between chemistry and biology. Most conspicuously, there was Francis Crick, a technical physicist who had been involved in problems of radar during the war, who together with James Watson in 1953 reconstructed the double helical structure of DNA from X-ray reflections. In the process, and this is what really made the discovery important, he concluded how genetic information could be stored and transferred from generation to generation. In Cambridge there was also Max Perutz, working in the Cavendish Laboratory under Sir Lawrence Bragg, whose method of X-ray interference patterns he applied to such complex molecules as the red blood cell dye, haemoglobin, elucidating together with John Kendrew for the first time the detailed design of a biomolecular machine. That was the birth of molecular biology.

Today we have a broad appreciation of the molecular design of living cells, including detailed mechanisms of the molecular processes lying at the basis of cell functions. We know about perturbations and breakdowns of such functions, as expressed in the most diverse sets of clinical symptoms; how parasites in the form of bacteria, fungi and viruses destroy the life cycle of an organism. Indeed, we can even go so far in regulating these life processes as to permanently alter their genetic program. Increasingly, the currently more chemically oriented pharmaceutical industry is exploiting our detailed knowledge of molecular biology and the associated technical opportunities. It is basic research, paramountly, that has irrevocably embraced the so-called recombinant DNA technology. What would we know about the molecular structures of the immune system, or about oncogenes or AIDs without this technology?

But I do not wish to bombard you with a quasi-alphabetical list of all the highlights of molecular biology, nor to confront you with a list of the names of those, from Avery, Luria and Delbrück to Neher and Sackmann, who excelled in creating them. Neither in my account do I want to deal with the biology of the first half of this century more specifically, other than to say that it was not just a completion of the grand concepts of the 19th century, of the ideas of Charles Darwin and Gregor Mendel, the insights of Louis Pasteur, Robert Koch, Emil von Behring and Paul Ehrlich. The first half-century established primarily a chemical foundation, through the work of Otto Warburg, Otto Meyerhof, his students Hans Krebs and Fritz Lipmann and many others, upon which the molecular biology of the second half-century could develop. I would much rather focus on the fundamental questions of biology. Answering them has only entered the realms of possibility through the compilation of detailed molecular knowledge in the 20th century. In doing so, we will cross the threshold into the 21st century and cast a glance into the future. Many questions that we can formulate today will only find a satisfactory answer in the coming century.

WHAT IS LIFE?

Not only is this a difficult question; perhaps it is not even the right question. Things we denote as 'living' have too heterogenous characteristics and capabilities for a common definition to give even an inkling of the variety contained within this term. It is precisely this fullness, variety and complexity that is one of the essential characteristics of life. Possibly it will not take very much longer until we know 'everything' about the *Escherichia coli* bacterium, perhaps even about the fruitfly *Drosophila*. But what will we then know about humans?

It is certainly then more sensible to ask: how does a living system differ from one that is not alive? When and how did this transition take place during the history of our planet or of the universe as a whole?

As a chemist I am often asked: what is the difference between a coupled chemical system albeit arbitrarily complex, and a living system in which we again find nothing other than an abundance of chemical reactions. The answer is that all reactions in a living system follow a controlled program operated from an information centre. The aim of this reaction program is the self-reproduction of all components of the system, including the duplication of the program itself, or more precisely of its material carrier. Each reproduction may be coupled with a minor modification of the program. The competitive growth of all modified systems enables a selective evaluation of their efficiency: 'To be or not to be, that is the question.'

There are three essential characteristics in this behaviour which are found in all living systems yet known:

1 Self-reproduction – without which the information would be lost after each generation.
2 Mutation – without which the information is 'unchangeable' and hence cannot even arise.
3 Metabolism – without which the system would regress to equilibrium, from which no further change is possible (as Erwin Schrödinger already rightly diagnosed in 1944).

A system that shows these properties is predestined to selection. I mean that selection is not an additional component to be activated from outside. It would be meaningless to ask who does the selecting. Selection is an inherent form of self-organization and as such, as we know today, a direct, physical consequence of error-prone self-reproduction far from equilibrium. Equilibration would only select the most stable structure. Selection – an alternative category incompatible with equilibrium – chose instead a sufficiently stable structure which is optimally adapted for certain functions which ensure the preservation and growth of the organism. Evolution on the basis of natural selection entails the generation of information.

In order to fix information structurally, defined classes of symbols are required, like the letters of an alphabet or the binary symbols of a computer code. Additionally, we need the connecting relations between symbols for forming words and the syntax rules which combine words into sentences. Facilities to read the sequences of symbols are admittedly also necessary and, ultimately, information is only that which may be understood and evaluated. The ability to deal with information in our language is coupled with the existence of a central nervous system.

What form does this take in the case of molecules? Information storage in molecules is subject to the same prerequisite that the information be 'readable' and subject to evaluation. Only with nucleic acids did molecules learn to read. Complementary interaction, an inherently specific association between two matching pairs of nucleic acid building blocks, underlies this ability of nucleic acids. So the basis of molecular information processing is base pairing, as discovered by Watson and Crick. This at first purely chemical interaction enables the transcending of chemistry, for the chemical building blocks act primarily as information symbols. Evolution, first molecular, then cellular and finally organismic, was only possible through reproduction and selection. It no longer selected according to purely chemical criteria but according to the functional encoding of information. Man differs from *E. coli* bacteria not in a more efficient chemistry but in greater information content (in fact a thousand times more than a *coli* bacterium). This information codes for sophisticated functions and makes complex behaviour possible.

The formation of a subcellular information processing system occurred 3.8±0.5 billion years ago, as we can reconstruct today from comparative studies on the adaptors of the genetic code. Accordingly, life probably began on earth and not just somewhere in the universe. It is not older, but also not much younger than our planet. This means that life arose as soon as conditions were suitable. There were already single celled organisms at least 3.5 billion years ago. Admittedly, the path to the true masterpieces of evolution, to the multicellular plants, insects, fish, birds and mammals, was a long and difficult one. It took all of 3 billion years. Mankind entered the stage in this magnificent drama only one million years ago.

Molecular biology has confirmed Darwin's fundamental idea through its ability to disclose what the genomes of living organisms have in common. Information, in this case genetic information, is formed by way of successive selection. Darwin proposed his principle for the evolution of autonomous living things. An extrapolation to precellular systems, to answer the questions 'How did the first life forms arise? From where did the first autonomous cell come?', seemed to him to be too daring a step. Once he did express a speculative 'if' and qualified it immediately with 'oh, what a big if!'. The exciting realization today is that selection is already active at the molecular level, with replicable molecules like RNA and DNA, and so is amenable to a derivation on the basis of the physico-chemical properties of molecules. This closes the gap which yawned between biology on the one side and physics and chemistry on the other. This does not imply that biology may be reduced to physics and chemistry in the conventional sense. It simply confirms that there is a continuity between physics, chemistry and biology. The physics of living systems has its own characteristic regularities. It is a physics of information production.

The new theory of self-organization goes far beyond Darwin in detail and answers questions that had to remain open or were even paradoxical in his time. Darwin's legacy is a testimony of the 19th century.

Ludwig Boltzmann once said (in 1886): 'If you ask me earnestly whether this century will be called the iron century, or the century of steam or of electricity, I must answer without hesitation, it will be called the century in which the mechanisms of nature were encaptured, the century of Darwin.' Surely Boltzmann hid his light a little under a bushel there in paying tribute to Darwin. Only today is it apparent that the reduction of living phenomena to a mechanical conception of nature is only one side of the story. The natural laws underlying selection and evolution overthrow any purely causal-mechanical conception of nature and describe a world with an open, indeterminable future. This change of paradigm, perhaps the only one in natural science which deserves the title, is not limited to biology. It has extended to the whole of physics over the past few decades and will work out its consequences over a far longer period. While learning how information can arise, we build a bridge between nature and mind.

HOW IS (BIOLOGICAL) INFORMATION GENERATED?

Since the middle of this century, we have been in possession of a theory bearing the name information theory. Its founder, Claude Shannon, however, pointed out from the beginning that it is not a theory dealing with information itself but rather with the communication of information. The information as such is excluded from consideration; it is treated as given: one sequence of symbols amongst many alternatives which must be maintained during transmission, irrespective of its semantic content or value. Information in this theory only features as a complexity measure. A string made up of two symbols, for example one and zero, of length N has 2^N possible alternative sequences. Even for relatively short sequences of length N about 300 (occupying a paragraph on less than half a printed page), the number of alternative paragraphs is larger than the number of atoms in the universe. Only a dynamic theory of selection can account for the difference between meaningful and meaningless sequences, by means of criteria which evaluate their semantic or phenotypic content. In order to enable an evolutionary optimization of this content, it must be reproduced with a finite error rate. Indeed there is an error threshold, immediately beneath which evolution is optimal, but above which the information falls victim to an error catastrophe. It vaporizes just as if at a material phase transition.

A modification of the Darwinian world view is already apparent here.

Natural selection is not simply an interplay between random mutation and deterministic, necessarily consistent selection. With such a large number of alternatives, the successful guesses for advantageous mutants would occur much too seldom. Today, this interplay of chance and necessity can be simulated readily on a computer. A process running according to this scheme is found to progress much too slowly. If natural selection had proceeded according to this scheme we would not exist.

In reality, molecular evolution near the error threshold involves an extremely broad spectrum of mutants. The best adapted (fittest) type, the wild type, which plays such a major role in Darwin's theory, is only present in relatively small numbers compared with the total population at the molecular level. However, the large number of mutants do indeed cluster about the best adapted type, so that the mean 'consensus' sequence does represent the entire population. Molecular biologists have learnt how to determine such sequences. Cloning experiments have revealed that the wild type does in fact correspond to the mean of a spectrum of myriad alternative sequences. In essence, this population is composed of only those mutants that can be efficiently reproduced. This theoretical result has been confirmed experimentally for virus populations. Since there are many billions of more or less mutated copies in such a molecular or viral distribution, which is fully stable below the error threshold, it is as if dice were being cast in a billion channels in parallel. If a better adapted mutant is found, the previous distribution is no longer below the error threshold. It becomes unstable and its information content vaporizes only to condense in the vicinity of a new wild type. Despite the continuity of the underlying molecular processes, we see that evolution proceeds via discrete jumps. Selection proves so efficient because it is a property of the whole population, representing a massively parallel sequence of events. If one wanted to simulate this process, one would need a new kind of parallel computer. To perform such a simulation on a serial computer would involve impractical demands on time and money. Nature demonstrates what form the computer of the future must take. Our brain is such a parallel computer with many billions of nerve cells, each one of which is connected with some 1000 to 10 000 neighbouring cells via synapses. Our immune system, too, is a cellular network of this order of complexity.

At the end of the 20th century, we are conscious that in many different branches of biology analogous questions are being formulated. These can be commonly phrased as 'How is information generated?' This is true for the process of evolution at the molecular level, for the process of differentiation at the cellular level and equally for the process of thought in a network of nerve cells. Still more exciting is the appreciation that nature apparently uses similar fundamental principles in quite different technical implementations in molecular genetics, the immune system and the central nervous system. The 1990s were designated in the USA as the decade of brain research. The legacy of biological research in this century will be a deep understanding of information-creating processes in the living world. Perhaps this entails an answer to the question 'What is life?'

Only, 'the devil is in the nuts and bolts'. Very soon, we will know the construction plans of many living organisms, and we will know how these have been found in the course of evolution. The historical roots, however, are still completely shrouded in mist. The scholastics once asked the question, what came first – the chicken or the egg, or, in modern terms, proteins or nucleic acids, function or information. The RNA world, containing as it does a genetic legislature and a functional executive, may offer a way out of this dilemma. I must admit that we do not (yet) know how the first RNA molecules 'entered the world'. From an historical perspective, the proteins should have come 'first', but historical precedence is not necessarily identical with causal precedence. Evolutionary optimization requires self-reproducing information storage and we only know nucleic acids to be capable of this role. So RNA, or a precursor, would have been necessary to set the merry-go-round of evolution in motion.

We are now in a position to observe in laboratory experiments the process of information generation in systems that contain both components: proteins (as enzymes) and nucleic acids (as information storage). Viruses are outstanding model systems. Viruses cannot however have formed in a pre-biotic world. They need a host cell to survive, with whose help they have evolved, that is, probably only post-biotically. Yet there is a strong analogy to virus-like RNA precursors in a host-like chemical environment.

The accumulation of knowledge about the process of information generation which we have achieved in the past 20 years is already beginning to bear fruit. Using laboratory methods, we will be able to produce new kinds of natural medicines and drugs. These skills are not restricted to the molecular level. In the same manner, we will understand the ontogenic level of living organisms and be able, for example, to intervene to heal tumours by causing them to degenerate. We will learn to know and to model our nervous system and its mode of operation. Artificial life and thinking computers will no longer be relegated to science fiction. It is scarcely possible to assess the impact all this will have on our lives.

But – there will be limits, both natural and normative. We will have to determine which parts of our knowledge we may apply, which parts we will have to apply despite an awareness of possible side effects and which aspects we must not meddle with, let alone apply. A blind frenzy of application is just as dangerous as a strict prohibition. We, the whole community of man, must find out rationally what should or should not be done, what must be done and what must not. Precisely in this context is where I see the biggest unresolved problem which will occupy us in the coming century.

WHAT PROBLEMS REMAIN UNRESOLVED AT THE END OF THIS CENTURY?

Some problems have been raised above; but even if I constructed a restricted list of only those problems which we can precisely define, it would be unmanageably lengthy. So I can only proceed by examples and I have chosen two problems from the heart of my own branch of research: one scientific problem with a great impact on society and a second problem where society has a great impact on science.

One problem which has not been solved despite the most intensive research is AIDS. What is AIDS? The word is an acronym for acquired immunodeficiency syndrome. The disease is initiated by a virus or, to put it more cautiously, is causally linked with a viral infection. The question whether the virus is both necessary and sufficient for the outbreak of disease symptoms is currently under vigorous debate. There are two known subtypes of the human immunodeficiency virus: HIV-1 and HIV-2. In addition, a larger number of monkey viruses have meanwhile been isolated that, while showing no pathogenic effect in their natural hosts, do so on transmission to other populations of monkeys. The US Center for Disease Control has established that on average ten years pass between viral infection and the outbreak of disease symptoms. More precisely, one finds that after ten years about 50 per cent of those infected show disease symptoms which quickly lead to a complete paralysis of the immune system. The disease AIDS then always results in death, mostly owing to an infection by a pathogen with which the immune system would normally have easily dealt. Many patients die of pneumonia, precipitated by a bacterium (*Mycobacterium tuberculosis*) which is latent in nearly every second person. During the symptom-free period, the AIDS virus itself is only present in a very small population in the organism. The latter produces antibodies in large quantities, with whose help the presence of the virus is detected in AIDS tests. In the USA, the number of registered cases of AIDS has now climbed to well over 100 000. Worldwide, the number of people infected with the AIDS virus is estimated to be nearly 10 million, with a concentration of cases in Central and West Africa and in South East Asia. No lasting therapy is known.

Where does AIDS come from? How old is the virus? When did it first appear in the human population? To answer these questions, the wildest hypotheses have been expressed. The ultimate was the claim that the virus was 'composed' in a US army laboratory and escaped by accident into the ecosphere. This is pure nonsense! Sequence analysis of the genes of this virus resolve its evolutionary history, or at least quantitatively constrain it. And these are the results:

- Both human subtypes, HIV-1 and HIV-2, as well as the currently known monkey viruses have a common ancestor that can be dated to about 1000 years ago.
- All HIV and SIV (simian immunodeficiency virus) sequences show matching positions (about 20%) and clear homologies in these to other mammalian retroviruses. The AIDS pathogen is thus the progeny of an old family of viruses whose origin reaches back many millions of years.
- The majority of the variable positions have a mean substitution time of approximately 1000 years. The special behaviour of the retrovirus, especially its pathogenicity, can change radically in such a period of time. Thus, plagues like AIDS may come and go. They may prove more pathogenic for some species and less so for others.
- A smaller portion (about 10%) of the positions prove to be hypervariable with a mean substitution time of about 30 years. This is, nonetheless, enough positions to generate an enormous number of different mutant combinations. Among these, escape mutants are repeatedly found that are not suppressed by the immune defence. In the end, this exhausts the immune system and is probably the main reason for the pathogenicity of the virus.
- The AIDS virus certainly did not appear in the USA, Europe or Japan before the 1960s. In Africa, related forms may be dated back to the previous century. During the last hundred years, horizontal transmissions

between monkeys and man are seen. The focal point of HIV-1 lies in Central Africa and that of HIV-2 in West Africa; HIV-1 and HIV-2 separated, like most species-specific monkey viruses, many hundreds of years ago.

The high pathogenicity of the virus has three causes:

1 Since HIV is a retrovirus, its genome is integrated in the genetic program of the host cell following infection. Once a cell is infected it can no longer free itself of the viral information. At most it can suppress its expression.
2 The target of the virus is the immune system itself, whose control centre is paralysed by the virus.
3 Because of its high mutation rate, which incidentally is right at the error threshold, the virus consists of a widely dispersed mutant spectrum containing a large number of escape mutants.

The virus evolves without pause under the selection pressure exerted by the host's immune system. The infected individual is eventually unprotected against some normally harmless parasite.

The difficulty in fighting the virus lies in its great adaptive potential. The virus manages to give the host's defence mechanisms the slip with the help of 'side-stepping' mutants. Since the viral strategy is now known, there is some prospect of finding an antiviral strategy that takes account of the side-stepping behaviour, leaving the virus no chance of surviving. To seek out such a strategy we need not only genetic technology but also animal experiments. Whatever our position on these, the reality is 10 million HIV-infected people, of whom the majority will develop AIDS symptoms by the turn of the century. Hardly any of these will survive – unless we have found an effective therapy by then.

My second problem has exactly the reverse polarity, directed from society to science. For several years now, we have had a gene law in Germany. Indeed it is the toughest in the whole world. It has begun to cause paralysis in research and industrial development. On the other hand, we should credit the fact that worldwide there have not yet been any mishaps or very serious accidents. Recent proposals even go so far as to require the prior proof of the absolute safety of a procedure. But what is 'absolute safety'? Even now, before any application of a procedure, every conceivable test is carried out and a long probationary period is adhered to. Nowadays the demand is being made to exclude things which are not yet known. This would bring research to a complete standstill and as a consequence make the development of new medicines impossible. (Proposals for animal protection laws also lead in this direction.) I will now give an example. Before the beginning of the 1960s, spinal paralysis of children, poliomyelitis, was a terrible plague in our latitudes. It manifested itself in both isolated cases and worldwide epidemics claiming many victims and many lifelong handicaps. In 1950 alone, 30 000 cases were recorded in the USA. Today, such cases have almost completely disappeared thanks to a rigorous program of prophylactic vaccination. Only in developing countries, and here largely because of an inadequate inoculation program, is poliomyelitis still a serious problem. The pathogen is a virus, a so-called picorna virus. There are at present two vaccines, either a mixture of killed virus (Salk vaccine) or a so-called 'attenuated' virus (Sabin vaccine) that is a mutant, no longer pathogenic, of the wild type that nonetheless evokes an immune reaction which is stronger than that of the killed virus. It is especially thanks to this orally dispensable vaccine, with its ease of use and great effectiveness, that the virus could be almost completely eradicated in the western world. Occasionally, instances of the illness are observed, but their course is relatively mild.

This was all very well. All the more unexpected was it when the RNA sequence of one of the Sabin vaccines (B-type) became known a few years ago. It turned out that essentially a two-error mutant of a pathogenic wild type was involved. Such a mutant can revert to the wild type inside 48 hours. Apparently, this period is long enough to initiate an effective response by the immune system. Since mutations are random events, occasionally more rapid back mutation is possible and this may be the cause of the isolated occurrences of the disease. In the case of AIDS, such a vaccination program would certainly set off a disastrous epidemic.

What is the difference between the polio virus and the AIDS virus? Both their genomes consist of a single RNA molecule. The mutation rates prove to be of a similar magnitude in both cases. With the help of a new method of comparative sequence analysis, called statistical geometry, we found out that there is a vast heterogeneity in the fixation of mutations in the different codon positions in the gene which codes for the surface proteins of the virus. Each protein building block is determined by a codon containing three positions. The first two determine the specific type of amino acid to be incorporated in the translation, while exchanges in the last position yield mostly synonymous amino acids, that is, producing no effect on the amino acid sequence of the translated protein. With the AIDS virus, all three codon positions are substituted at an equally high mutation rate, thereby creating a large spectrum of different protein molecules (some of which escape from the immune response). However, in the case of the polio virus, almost the only mutations that are fixed are those in the third codon position. At least, virtually all of the different, widely spread mutants that have been found differ only at the third codon positions. These substitutions are so numerous that there is almost complete substitution, while the first and second codon positions remain almost unchanged in all mutants. This means that the proteins on the surface of the polio virus scarcely change. There are hence no escape mutants. The immune system can 'get its eye in'; that is, an effective immune defence is built up within a short period.

But now to the moral of the story: had one known that the attenuated virus was such a close relative of the pathogenic wild type, there would surely have been the gravest scruples about allowing such a vaccine. According to the current view, this would certainly not be possible since it has become common to produce such mutants through directed mutagenesis, that is, by genetic engineering. One would now be 'circulating a genetically manipulated pathogenic agent'. At any rate, we could not exclude with our current state of knowledge a risk that is in fact a real one, as shown by the occasional incidences of the disease after oral vaccination. Nothing was known of all this when the Sabin vaccine was introduced. One proceeded, quite legitimately, by empirically testing the attenuated virus. There was then absolutely no other alternative.

However, a genetically engineered mutant is no different from one arising naturally. In one case, we manipulate and *we know what happens*. In the other, Nature manipulates by herself and we do not know what comes out but can only test empirically what happens. One method is branded evil, the other accepted as natural, although it is always easier to control a risk by *conscious* action than by unconscious tinkering. Reading the text of the German gene law, one repeatedly meets such nonsense; one wishes to exclude every risk 100% while accepting other imponderables without consideration. For example, research work that could one day serve to ward off a danger is totally suppressed. In the case of polio, one would certainly have avoided the not quite risk-free path of genetic manipulation and that would have *de facto* meant the death of many children. The Sabin vaccine saved them, because one trusted nature blindly and thereby accepted unconsciously the inherent risk.

In this context, the question must be raised: how far should the indifferent majority of society give in to the ideological arguments of an emotionally aroused minority against the advice of specialists? What is, ultimately, freedom of research, as guaranteed in the German constitution? I wish by no means to interpret freedom as a total lack of restraint. We can neither do all that we know nor should we do all that we are able. What other way is there to make decisions if not rationally? In the case of Hiroshima, there was insufficient military and political sense, and in the case of Chernobyl, there was too little technical sense. Knowledge cannot be 'undiscovered'. We must learn to live with it. For this purpose, a sensible legal structure is required that should be *internationally* binding. Over and above this, there is an ethical duty to employ available knowledge for the benefit of mankind, whether it be to reduce the suffering of individuals or to ensure the health and nourishment of the world's population. I return to the scenario of the future of mankind which I described in my introduction. An environmentally just safeguarding of food production for a multi-billion strong world population, an adequate system of sanitary and medical care for such 'masses of people': these are things only possible today if all available knowledge is applied. This includes a genetically engineered breeding of new organisms for food as well as the use of nuclear technology for electricity production.

THE FUTURE: THE STUDY OF MANKIND IS MAN!

We live in a society that shies away from risk. Will it come to the point that society, for this very reason, closes the door on science and especially on basic research. It would not surprise me even now to see on the rear window of a car, belching blue-grey exhaust, the sticker 'Basic research – no thank you!'. What some members of the animal protection movement are engaged in is pitched at least down at this level. The opponents of atomic power are happy that electricity flows from the wall sockets at home. We can do nothing useful without taking on risks at the same time. Failing to do anything can be the greater evil in the long run. We must learn to weigh odds, and slogans are not very helpful in this.

When I speak of the future of biological research, it will be of increasing problems of risk evaluation, of responsibility and ethics that we must debate. For the central object of biological research is man and his environment, 'his' meaning relative to man. Consequently, the results of resear relevant to everyone.

I do not want to attempt to construct scenarios for the coming century, let alone for the coming millennium. According to Friedrich Dürrenmatt, problems are only fully thought out 'when one has imagined the worst possible turn of events'. Indeed, futurologists are liable to paint only the rosiest possibility.

We will be able to explore the genetic nature of man far better than we have ever dreamed for there will be machines that will be able to read the three billion letters of human inheritance inside a month. This will, in particular, enable comparative studies to be made. In the same way, we will determine the genetic sequence of very many other life forms and then be able to unravel our own evolutionary origin. We will fathom the human brain and build computers that far exceed the brain in particular tasks. I do not believe that we will ever possess a computer that even approximates the human brain in all its capacities, but a connected brain and computer will demonstrate 'superhuman' capabilities. We will not be able to crystallize an homunculus, but robots will be invested with powers hitherto met only in the biological realm. Whether we call this 'Artificial Life' or not is only a matter of taste. We will be able to cure cancer, because we are uncovering more and more of its causes. Furthermore, for heart disease, we will be in a position to make earlier diagnosis allowing medical assistance to come in time. Nevertheless, in the long run it will be immaterial from which illness we die, because even in the future I expect our age to hardly exceed 100 years. We need not actually worry whether the cities of the future will have a glass dome and an artificial atmosphere. But we certainly should ask ourselves today: where will we obtain all the energy that will be needed one day to maintain a recycling economy? Keeping air and water clean is a task bound up with high entropy production. A timely provision for the future is essential here. Admittedly, there will be many novel discoveries and inventions that defy our imagination now. It is just for this reason that every detailed scenario for the future will be wrong. We are in the same position as Charlemagne would be if his contemporaries had asked him about the world of the 20th century.

Despite this, one prognosis is reasonably certain: whether the course of humanity takes the worst or the best possible turn will depend on whether man finally learns what he has failed to learn in the five millennia of his cultural history, namely, to act rationally and sensibly in the interests of humanity and to work out well-defined rules of conduct. The latter are analogous to a genetic program and must be established as binding for all.

Man stands on the highest rung of the ladder of evolution. I say this, not because I cannot imagine any creature more perfect, but because with man evolution has reached a new platform, accessible to no other organism, from which evolution must proceed in a radically new fashion. Operating on the basis of selection, evolution requires the continuous mutagenic reproduction of information, laid down like printers' type in our genes. New avenues of communication have opened up between cells with the formation of cellular structures and networks. These were originally mediated by chemical signals that are collected by specific receptors and ultimately mediated by electrical signals that are received by synapses and relayed to the next cell. By this means, a correlated overall behaviour of the differentiated cell system could develop, preprogrammed in the genome only in its layout. It is selection that ensures that this layout operates to the advantage of the whole organism. This is incompatible with single cells or organs working against one another. Such antagonism can only take the form of diseased degenerations like cancer. In the central nervous system, this intercellular communication has developed into an inner language that controls our behaviour, emotions, disposition and feelings. Even this facility has become genetically anchored and has been selected so that it is not directed against the species. This is the way in which man arose in the course of evolution and this genetically programmed, individualistic and species-specific behaviour is inherently egoistic, set on competition and self-assertion. In places where instead it appears altruistic, it rather serves in the long run the advantage of the species or clan, which, in its turn, has an advantageous effect in some way on the individual.

Man has, by this means, developed a specific faculty, distinct from that of other primates, that allows him to achieve a formalizing of the inner language coded primarily in nerve cell discharges. This formalization not only facilitated the communication between members of the species, it also formed the basis of our ability to think, to record results for the benefits of mankind and to bequeath them to following generations by setting them down in writing. This implies a new plane of information transfer, similar to the primary plane of genetic information which added a totally new quality to chemistry. On the plane of the human mind, a new form of evolution can take place: the cultural evolution of humanity.

However, there lies the key problem. Humanity is not something like a multicellular organism in which every cell leads its individual life but is committed by the genetic legislature to the common good of the cell community. Cultural information is not inherited by the individual, just as little as is socially acceptable behaviour. Despite the cultural evolution of humanity, which has lasted now for thousands of years, people still engage in war and no less cruelly than ever. We delude ourselves if we believe that socially acceptable behaviour is something natural and asocial behaviour, in contrast, something pathological. It is the norm only in the original sense of the Latin word *norma*, meaning rule or regulation.

We find ourselves in a genuine dilemma, for all previous attempts to subject individual freedom to dictates, degenerating the individual to the status of a cell without will in a centrally controlled organic whole, have only harmed human society in the long run and have even resulted in the annihilation of parts of humanity. These experiments have failed, partly because the new organism was not the whole of humanity but only a certain grouping, representing special interests which often violated basic human rights. Partly they have failed because the 'leading cells', the 'brain cells' of this large organism, were mostly self-obsessed or egoistic human cripples, primarily concerned with exercising power. Incomparable suffering was the result.

Ideologies cannot replace reason. All political groupings that advocate party discipline should come to appreciate this. Of course, they stand for ideals that have a valid basis, whether they call themselves Socialists – who would not be for a social conscience? – or Greens – who would not like to keep the environment clean? – or Christians – who would wish for a world without mercy or charity? This holds equally for those who want to place the freedom of the individual above everything else. Each of these motives, raised alone to a pedestal of doctrine, is directed against our common sense, in which, incidentally, not only our intellect but also our limbic system, our feelings and emotions are involved. Even in the future, we will by no means be able to delegate our judgement to a computer.

One glance at the current state of the world is apt to make us pessimistic. The first half of this century has dealt us two of the most shocking wars. And what lessons have we learnt? Nothing will change if we do not base our decisions on reason, accepting humanity as a moral imperative. The future of mankind will not be decided at the genetic level. We need a binding system of ethics for all people. Here evolution, an evolution from the individual to humanity, awaits its consummation.

3

'What is life?' as a problem in history

STEPHEN JAY GOULD
Museum of Comparative Zoology, Harvard University, Cambridge, Massachusetts

WHAT IS LIFE? AS A MODERNIST MANIFESTO

The obviously true may be devilishly difficult to define – as best exemplified by Louis Armstrong's famous retort to a naively passionate fan's request for a definition of jazz: 'Man, if you gotta ask you'll never know.' It is similarly undeniable that Erwin Schrödinger's *What is Life?* ranks among the most important books in 20th century biology, but the reasons for its great influence seem oddly elusive. Brevity may be the soul of wit (as garrulous old Polonius told us), and short works are rare blessings in a profession that too often judges worth by literal ponderousness. But *What is Life?*, in its ninety pages, seems a bit too spare and too elliptical to carry such intellectual weight (though, in a ruthlessly practical sense, such brevity may define the essential differences between attention and oblivion in a profession dominated by doers rather than readers). For example, I think we may be confident of the correct, if necessarily conjectural answer to an old puzzle in 'iffy' history: how would the history of science have differed if Wallace had never lived and Darwin had thereby acquired the leisure to write the many-volumed work he intended, rather than the hurried 'abstract' known as the *Origin of Species?* The answer – since the intellectual world was clearly poised to accept evolution – must be: none whatsoever, except that Darwin would have had the same impact with many, many fewer people having read the book. Moreover, much of *What is Life?*'s intellectual foundation – Delbrück's early ideas on reasons for the gene's stability – turn out to be quite wrong (see Crow, 1992, p. 238). Why, then, are we so rightly celebrating this semicentenary?

First of all, the testimony of seminal importance by so many of the founders of modern molecular biology cannot be gainsaid. Jim Watson credits

Schrödinger's book as the decisive influence in persuading him to study the structure of the gene (see Judson, 1979). Francis Crick acknowledges a similar influence, but with the same puzzlement that so many others express: 'It's a book written by a physicist who doesn't know any chemistry. But . . . it suggested that biological problems could be thought about, in physical terms – and thus it gave the impression that exciting things in this field were not far off' (quoted in Judson, 1979, p. 109). (On the subject of puzzlement, consider Jim Crow's recent comment (1992, p. 238): 'Along with Gunther Stent, I don't know why the book had such an impact, but I do know what most impressed me at the time.')

Crow then provides an excellent epitome of the book's chief claims and insights – the second reason for its influence:

> Perhaps it was Schrödinger's characterization of the gene as an 'aperiodic crystal'. Perhaps it was his view of the chromosome as a message written in code. Perhaps it was his statement of life 'feeds on negative entropy'. Perhaps it was his notion that quantum indeterminacy at the gene level is converted by cell multiplication into molar determinacy. Perhaps it was his emphasis on the stability of the gene and its ability to perpetuate order. Perhaps it was his faith that the all too obvious difficulties of interpreting life by physical principles need not imply that some super-physical law is required, although some new physical laws might be.

I do not desire to denigrate this timely celebration by denying in any way the importance of *What is Life?*, but I do wish to suggest that Schrödinger's key claim for an almost self-evident universality in his approach to biology is both logically overextended, and socially conditioned as a product of his age. Furthermore, these features of limitation may help us to understand why a large subcommunity of biologists, including my own *confrères* in palaeontology and evolutionary studies, have been less influenced and impressed by Schrödinger's arguments, and remain persuaded that the answer to 'what is life?' requires attention to more things on earth than are dreamed of in Schrödinger's philosophy.

Schrödinger (1944, p. vii) begins his preface by identifying unification as the unquestioned dream and goal of science:

> We have inherited from our forefathers the keen longing for unified, all-embracing knowledge. The very name given to the highest institutions of learning reminds us, that from antiquity and throughout many centuries, the universal aspect has been the only one to be given full credit . . . We feel clearly that we are only now beginning to acquire reliable material for welding together the sum-total of all that is known into a whole.

Schrödinger presents this goal of unification as the unquestioned, almost logically necessary, yearning of all scientists in all ages. Quite the opposite is true. Unification was a definite aim of an explicit movement, embedded in particular social circumstances of Schrödinger's young adulthood; hopes for rational universality following the nationalistic carnage of the First World War. When we grasp the socially contingent character of this cardinal belief in unification, we can understand why Schrödinger's answer to 'what is life?' does not have general status, but must be seen as a transient product of one phase in 20th century history.

The self-styled 'unity of science movement' arose as a major aspect of logical positivism, developed by the Vienna School of philosophers during the 1920s. Associated primarily with Rudolf Carnap and Otto Neurath, both leading members of the *Wiener Kreis*, the unity of science movement held that all sciences share the same language, laws and methods, and that no fundamental differences exist between the physical and biological sciences, or (for that matter) between the natural and social sciences properly constituted.

The unity of science movement had great influence in biology, a field previously viewed by many as too idiosyncratic or descriptive to fall under the umbrella of generalized scientific theory (see Smocovitis, 1992, on the role of this doctrine in the evolutionary synthesis of the 1930s and 1940s). Schrödinger occupied an ideal position for translating the goals of this movement into biology. He was born and raised in Vienna and matriculated at the University of Vienna. He won a Nobel prize in physics – the 'focal' or 'highest level' science, towards which all others would be gathered in the fundamentally reductionist view of the unity of science movement, and of logical positivism in general. How could Schrödinger not anchor his book in a search for unification based upon physical laws?

If Schrödinger's belief in reductive unification flowed from the unity of science movement, then this movement, and its philosophical basis, also lay embedded within the even larger cultural force later known as 'modernism', with its profound influence upon such fields as art, literature, and architecture. Modernism, above all, sought reduction, simplification, abstraction, and universalism. In the hands of a master, like the architect Mies van der Rohe, modernist buildings (of the 'international style', named for the goal of universalism) may be elegant and powerful; but the thousands of derivative, substandard knock-offs, now cracking and deteriorating all over our planet, are the blight of Third World cities and the antitheses of legitimate regionalism and local pride.

What is Life? has usually been viewed as a timeless statement about the unchanging logic of science; I suggest an opposite reading as a social document representing the aims of the 'unity of science movement', as an expression of the larger worldview known as modernism. As such, the faults and strengths of Schrödinger's book are tied to the failures and successes of modernism in general. I can applaud much of modernism's spirit, particularly its optimism and its commitment to mutual intelligibility based on unity of principles. But I also deplore its emphasis upon standardization in a world of such beautiful diversity; and I reject the reductionism that underlies its search for general laws of highest abstraction.

In our generation, these widely recognized social faults of modernism (particularly its tendency to award hegemony to one fashion over other legitimate contenders) has spawned a countermovement known (with no great imagination) as 'postmodernism'. And whereas I regard much about postmodernism as rueful (from silliness in architecture to opacity in literature); and while postmodern 'improvements' must be viewed not as higher truths but as social signs of our own times (just as modernism reflected earlier decades), I nonetheless find much of enormous value in the general postmodernist rejection of modernism's search for single, abstract solutions. I particularly applaud postmodernism's emphasis on playfulness and pluralism, its approbation for the irreducible importance of local detail, and its conviction that, although truth itself may be unitary (many postmodernists would deny this claim as well, but here I part company from such tendencies to nihilism), our perspectives upon truth may be as multiply valid as our socially embedded modes of seeing. A postmodernist could scarcely credit any unitary answer to such a question as 'what is life?' – particularly an answer, like Schrödinger's, rooted in the modernist heartland of reduction to constituent basic particles.

In short, I love much of Schrödinger's book, while I regard its faults as expressions of general problems with the philosophy of modernism that permeates the work. As an evolutionary biologist committed to the study of whole organisms and their histories, I do not regard Schrödinger's answer as wrong, but only woefully partial, and scarcely touching some of the deepest issues in my field.

One could hardly propose a more congenial or more conciliatory form of reductionism than the argument advanced by Schrödinger as the centrepiece of *What is Life?* – for he does not advance the haughty old Newtonian claim that biological beings are 'nothing but' physical objects of high complexity, and therefore ultimately reducible to conventional concepts developed by the queen of sciences. Schrödinger admits that biological objects are different and unique. They must ultimately be explainable by physical principles, but not necessarily the ones we know already. Therefore, biology will be as much help to physics (in providing material that will lead to discovery of these unknown laws), as physics can be to biology in finally supplying a unified explanation for all matters:

From Delbrück's general picture of the hereditary substance it emerges that living matter, while not eluding the 'laws of physics' as established up to date, is likely to involve 'other laws of physics' hitherto unknown, which, however, once they have been revealed, will form just as integral a part of this science as the former. (Schrödinger, 1944, p. 69.)

Schrödinger then tries to deduce the nature of hereditary material from its failure to operate by physical laws known to apply to the smallest particles of nonliving matter:

From all we have learnt about the structure of living matter, we must be prepared to find it working in a manner that cannot be reduced to the ordinary laws of physics. And that not on the ground that there is any 'new force' or what not, directing the behaviour of single atoms within a living organism, but because the construction is different from anything we have yet tested in the physical laboratory. (Schrödinger, 1944, p. 76.)

In his new quantum world, the 'probability mechanism of physics' (Schrödinger, 1944, p. 79) builds macroscopic order from molecular disorder – 'our beautiful statistical theory of which we were so justly proud because it allowed us to look behind the curtain, to watch the magnificent order or exact physical law coming forth from atomic and molecular disorder' (Schrödinger, 1944, p. 80). The complexity of hereditary material will require a new principle of order from order:

The orderliness encountered in the unfolding of life springs from a different source. It appears that there are two different 'mechanisms' by which orderly events can be produced: the 'statistical mechanism' which produces 'order from disorder' and the new one, producing 'order from order' . . . Physicists were so proud to have fallen in with . . . the 'order-from-disorder' principle, which is actually followed in Nature . . . But we cannot expect that the 'laws of physics' derived from it suffice straightaway to explain the behaviour of living matter, whose most striking features are visibly based to a large extent on the 'order-from-order' principle. You would not expect two entirely different mechanisms to bring about the same type of law – you would not expect your latch-key to open your neighbour's door as well. (Schrödinger, 1944, p. 80.)

These arguments led Schrödinger to his most famous inference, the one that secured such historical influence for his small book – the concept of the gene as an 'aperiodic crystal'.

'WHAT IS LIFE?' A QUESTION FOR PLURALISM

A titular problem

Given the context presented above, I trust that I shall not be deemed either too carping or too trivial if I state that my major problem with *What is Life?* resides in the implied claim of its title. Right on page one, Schrödinger states the question that his book will try to answer:

The large and important and very much discussed question is: How can the events in space and time which take place within the spatial boundary of a living organism be accounted for by physics and chemistry? (Schrödinger, 1944, p. 1).

(This formulation at least provides a stage as broad as an entire living creature, although *What is Life?* then goes on to discuss, almost exclusively, the physical nature of the hereditary material.)

In short, and in the spirit of reductive modernism, Schrödinger argues that we shall have our answer to 'what is life?' when we know what the smallest pieces of heredity are made of, and how they work in a universal mode. I do not deny the inestimable value of learning the nature and construction of genetic material. But does this knowledge give us an adequate answer to 'what is life?' Is there not more, very much more, that any sensible vernacular concept of such a question must include? From a purely parochial standpoint as a palaeontologist, I must reject Schrödinger's narrow formulation, for its acceptance makes my field irrelevant or, at best, entirely subsidiary. If knowing the physical nature of hereditary material answers 'what is life?' then why is my profession trying so hard to trace phyletic history on the grand scale of billions of years? At best, the earth could only be a stage for documenting details of a history specified by a theory developed entirely from understanding the nature of constituent matter in its smallest pieces. In this view, palaeontologists have no theory to develop from their macroworld, no constituent to supply towards a full answer of 'what is life?' We can only document an actual, realized history, and such an activity becomes trivial if no theoretical insight can arise thereby.

Indigenous sources of reduction

What is life, then, beyond the working of its littlest pieces? Why did we ever think that we could adequately answer such a far-reaching question within such a restricted domain? – and why were so many of us fully satisfied with partial answers like Schrödinger's? In part, the blame lies in a series of traditions and social factors external to palaeontology and other subdisciplines of 'whole-organism' biology. Physics envy made the proclamations of great scientists in this field, particularly by Nobel laureates (for our disciplines are not honoured at all by such prizes), worthy of special respect (and, to a large extent, immune from searching criticism). The popularity of modernism gave undue clout to old reductionist foibles. Lack of sufficient pride in our own material (another consequence of reductionism and physics envy) made us more receptive to gurus from elsewhere.

But another set of factors arises from our own traditions and conventional forms of explanation – and we therefore have only ourselves to blame for too ready an acceptance of reductionism, and too willing an abandonment of our own phenomena as a rich source of theory for large parts of a full answer to 'what is life?' Classical Darwinism itself not only accepts but actively promulgates a style of reductionistic thinking that had rendered the geological stage theoretically irrelevant even before molecular genetics supplied an even harder version.

Two features of the strict Darwinian worldview encourage reduction of the geological pageant of life's history at least to the momentary machinations of organisms, if not right down to the physicochemical nature of genetic material. First, the theory of natural selection identifies a unitary locus of causal change as the organism struggling for reproductive success – and explicitly denies active causal status to any 'higher' biological unit like species or ecosystems. The beauty and radicalism of Darwin's system lies largely in its denial of overarching ordering principles (like God's action in older theories), and its attribution of higher-order phenomenology (like the harmony of ecosystems or the good design of organic architecture), to consequences or spin-offs of lower-level causality.

Second, under the grand vision of uniformity, as preached so effectively by Darwin's guru Charles Lyell, all scales of time, and all magnitudes of events, flow smoothly upward as summations and extrapolations from observable causal happenings of minimal effect occurring in moments – the Grand Canyon as accumulated erosion, grain by grain, over millions of years; evolutionary trends as gradualistic accretions of minimal changes, generation after countless generation.

We note this causal smoothness from the smallest scales in Darwin's own construction of natural selection as an analogue to the observable, even smaller-scale processes of artificial selection in domestication and agriculture. If humans, with such imperfect knowledge, have wrought changes over centuries, think what a ruthlessly efficient nature can do in extrapolated vastness:

> As man can produce and certainly has produced a great result by his methodical and unconscious means of selection, what may not nature effect? Man can act only on external and visible characters: nature cares nothing for appearances . . . She can act on every internal organ, on every shade of constitutional difference, on the whole machinery of life . . . How fleeting are the wishes and efforts of man! How short his time! and consequently how poor will his products be compared with those accumulated by nature during whole geological periods. (Darwin, 1859, p. 84.)

Moreoever, nature's stage plays out little daily events into any needed magnitude through the simple agency of sufficient time. We need no new forces for larger scales, no catastrophes of global proportions. Reductionism works because the entire causal structure for the history of both earth and life lies fully exposed in minimal events of observable moments.

This belief in causal uniformity established a gradualistic credo responsible for a range of fallacies in our understanding of natural history – from comforting iconographies (see Gould, 1989) of life's history as a ladder of progress (for morphology) or a cone of increasing breadth (for diversity), to dogmas about the steady course of geological change, as well captured in the opening of Davies's recent review of Derek Ager's posthumous book on neocatastrophism:

> 'Fascist!' Among street politicians that is the ultimate abuse shouted as a prelude to yet more violent leftist action. 'Catastrophist!' In my early days that was the ultimate insult to be hurled at an Earth scientist who seemed to be straying outside the prevailing dogma of uniformitarianism . . . We preferred to believe that what was important in geohistory was nature's long-term gradualistic processes . . . Sedimentary strata

formed in a marine environment were interpreted as the little-by-little accumulation of particles raining down on the sea bed over aeons of time. (Davies, 1993, p. 115.)

'WHAT IS LIFE?' AS A PROBLEM IN HIERARCHY AND HISTORY

In the pluralistic spirit of postmodernism, contemporary evolutionary theory is now moving away from the restrictive reductionism both of Schrödinger's sort (that 'what is life?' may be answered by knowing the physical nature of the smallest constituent parts) and of Darwin's (that higher-level processes and time scales can be explained as causal extrapolations from processes operating on individual organisms in the observable present). Two themes, hierarchy and the contingency of history, help us to realize that resolution at Schrödinger's or Darwin's level provides only partial answers to 'what is life?', and that many vital and legitimate questions contained in this conundrum of the ages require a body of theory – not just a phenomenology – operating at, and only extractable from, processes of time's macroscale and evolution's major transformations.

Hierarchy

Two separate themes, based on the general concept of levels of organization in times and magnitudes, preclude an adequate resolution of 'what is life?' at the scale of genes and their construction.

Hierarchy in the formulation of an evolutionary theory of selection. A kind of descriptive hierarchy was always acknowledged by the founders of modern evolutionary theory (see Dobzhansky, 1937 and the commentary in Gould, 1982), but these scientists generally accepted a causal reduction to changing gene frequencies within populations. Proposals for an explicit causal hierarchy within selection theory have inspired a major debate since the early 1970s. The mildest form of hierarchy holds that events of macroevolution, though fully consistent with microevolutionary theory, could not be predicted from tenets of the microworld and therefore demand direct attention to the phenomena of larger scale (Stebbins & Ayala, 1981).

The stronger form departs from Darwin's key claim that organisms are the exclusive locus of natural selection (or from the even more reductionist argument of Dawkins (1976) and others that genes may serve as such ultimate 'persons'). The hierarchical theory of natural selection holds that biological objects at several ascending levels in a structural hierarchy of inclusion – genes, organisms, and species prominently among them – may all act (simultaneously) as legitimate foci of natural selection. (Species are natural objects, not abstractions, and they maintain all the key properties – individuality, reproduction, and heredity – that permit a biological entity to act as a unit of selection.) If species are important units of selection in their own right, and if much of evolution must be understood as differential selective success of species rather than extrapolated predominance of favoured genes in populations, then evolutionary pattern – an important component of 'what is life?' – must be studied in the fullness of species durations, that is, directly in geological time (see Stanley, 1975; Vrba & Gould, 1986; Lloyd & Gould, 1993; Williams, 1992).

The earth's behaviour. Even if natural selection could, in principle, build evolution at all scales by simple cumulation, the earth must behave in a congenial manner to permit the gradualist throughput. If the earth be so unruly that slowly accumulating sequences are derailed or reset by occasional catastrophes of major import, then the causes of overall evolutionary pattern are complex – and the component attributable to rare occurrences of great moment cannot be grasped by traditional uniformitarian study of ordinary current events.

The virtual proof (Krogh *et al.*, 1993) of the Alvarez hypotheses of mass extinction by meteoritic impact at the close of the Cretaceous period (Alvarez *et al.*, 1980) has fuelled a general reexamination and willingness to admit an important role for such events and processes at higher levels in hierarchies of times and magnitudes. Davies (1993, p. 115) continues his critique of classical uniformitarianism:

> Now all is changed. We are rewriting geohistory. Where once we saw a smooth conveyor belt, we now see a stepped escalator. Upon that escalator the treads are long periods of relative quiescence when little happens. The risers are episodes of relatively sudden change when the landscape and its inhabitants are translated into some fresh state. Even the most staid of modern geologists are invoking sedimentary surges, explosive phases of organic evolution, volcanic blackouts, continental collisions, and terrifying meteoroid impacts. We live in an age of neocatastrophism.

Consider just three examples of macroevolutionary phenomena, all much discussed during the past twenty years, that must constitute a major part of any satisfying answer to 'what is life?' yet cannot be adequately resolved by understanding the construction of genetic material, or by any sensible extrapolation from this microlevel alone. (1) Evolutionary trends in a world of punctuated equilibrium (Eldredge & Gould, 1972; Gould & Eldredge, 1993), where directionality results from differential success of biased subsets of stable species within clades, and not from anagenetic transformation within lineages, and where a substantial component of differential species success occurs by irreducible selection at the species level itself. (2) Mass extinctions that are more rapid (some triggered by true catastrophes on scales of moments to days with main killing effects perhaps spreading only to centuries or millennia), more profound in effect, more frequent in occurrence, and more different in causality than we had ever imagined in our previously favoured Lyellian mode. (3) The restriction in time and magnification in effect for episodes of origin in life's history, particularly for the 'Cambrian explosion' that initiated virtually all the major designs of multicellular life. The Cambrian explosion has now been restricted by new and rigorous radiogenic age dates to a period of only 5 million years or so (Bowring *et al.*, 1993). Contrary to previous, conventionally progressivist, views that only the precursors of modern forms arose in this event, a thirty year restudy of the Burgess Shale (the spectacular, soft-bodied fauna from the Middle Cambrian period, just following the explosion) suggests that the range of these initial anatomical designs exceeded modern boundaries (despite more than 500 million years of subsequent time to generate new anatomies), and that the history of life since the Cambrian explosion has been largely a story of reduction in initial possibilities. With one exception (Bryozoa at the beginning of the subsequent Ordovician period), no new phylum has arisen in the fossil record since the Cambrian explosion. Whatever genetic and developmental setting permitted this cardinal event, it was not business as usual, to be simply extrapolated from Darwinian changes in modern populations (see Whittington, 1985; Gould, 1989). We cannot begin to answer 'what is (multicellular) life?' without understanding such events.

History's contingency

Apply all the conventional 'laws of nature' type explanations you wish; add to this panoply all that we will learn when we grasp the laws and principles of higher levels, greater magnitudes and longer times – and we will still be missing a fundamental piece of 'what is life?' The events of our complex natural world may be divided into two broad realms – repeatable and predictable incidents of sufficient generality to be explained as consequences of natural law, and uniquely contingent events that occur, in a world full of both chaos and genuine ontological randomness as well, because complex historical narratives happened to unfurl along the pathway actually followed, rather than along any of the myriad equally plausible alternatives.

These contingent events, although mistrusted and downgraded by traditional science, should be embraced as equally meaningful, equally portentous, equally interesting, and even equally resolvable as the more conventional predictabilities. Contingent events are indeed unpredictable, but this property flows from the character of the world – and becomes thereby as immediately meaningful as anything else presented by nature – and not from limitations of our methodologies. Contingent events, though unpredictable at the onset of a sequence, are as explainable as any other phenomenon after they occur. The explanations, as contingent rather than law-based, do require a knowledge of the particular historical sequence that generated the result, for such resolutions must be in the narrative rather than deductive mode. But many natural sciences, including my own of palaeontology, are historical in this sense, and can provide such information if the preserved archive be sufficiently rich.

A downgrader of contingency might admit all the foregoing claims, and still respond: yes, I grant your two realms, but science is only about the 'upper' domain of generality. The 'lower' region of contingency is small and flat, weighted down by the grandness above, and only the space of funny little details that have no importance in nature's basic working. The key to my argument lies in the denial of this common conceptualization, and in the restructuring of contingency's domain as equally broad and important as anything deducible from natural law – for contingency's domain embraces questions of the common form: 'why this, and not any one of a thousand something elses?'

The major argument may best be put as a historical or psychological observation. In our arrogance, but also in our appropriate awe, we tend to pose our deepest biological questions as generalities to be resolved by natural law: why must life run by natural selection on substrates built from nucleic acid codes? What in ecological theory will tell us why the earth houses so many insects and so few pogonophorans? What, after all, is life? (as a predictable phenomenon that would evolve again in the same manner and cannot be much other than it is). Yet most of these questions arise because we want so desperately to understand something just as puzzling, and much more particular: who are we as human beings, and why are we here? Protagoras was right in his famous aphorism that 'man is the measure of all things' (to be read either as a statement of ultimate humanism or as a narrow parochial claim). Now we, as a single species, the end product of a contingent sequence that could never have led to anything like us if any among thousands of preceding steps had unfolded even slightly differently (as each plausibly might have done) – we who are contingent entities, not predictable inevitabilities – lie firmly within the domain of contingency. And questions that are truly and deeply about us in particular, even if conventionally framed as inquiries about timeless essentials, are inquiries to be answered in terms of contingency.

Tiny differences in the realm of contingent history, seemingly inconsequential to any observer at the time, cascade to utterly disparate outcomes that fundamentally alter 'what is life?' Contingency is not the domain of trivial things alone. Contingency's theme, moreover, is fractal, and pervades all scales of life's history from biospheric cataclysms to particularities of single lineages. Why is *Homo sapiens* here? – the question that truly prompts our inquiry into 'what is life?' (as we will admit in honest moments). Go down the fractal scales and find contingency throughout. We are here because the death roster of anatomical products of the Cambrian explosion did not include a small and 'unpromising' chordate group represented in the Burgess Shale by the genus *Pikaia*. (Any replay of life's tape through the Burgess lottery would have yielded an entirely different cast of surviving lineages; in this sense, any group alive today owes its existence to contingent fortune.) Step down to the survival of mammals. No late-Cretaceous bolide (the ultimate random bolt from the blue), and dinosaurs would still be dominating the world of terrestrial vertebrates, with mammals probably still restricted to rat-sized creatures in the interstices of their world (dinosaurs had so dominated mammals for more than 100 million preceding years so why not for an additional 65 million?). Step down to a lineage of apes 10 million years ago in African forests. On this replay, climatic drying does not occur, forests do not convert to savannas and grasslands. The lineage stays in the persistent forest as apes – doing quite well in an alternate today, thank you.

Schrödinger wrote of his formative likes and dislikes: 'I was a good student, regardless of the subject. I liked mathematics and physics, but also the rigorous logic of the ancient grammars. I hated only memorizing 'chance' historical and biogeographical dates and facts.' How ironic that a great pioneer in a scientific revolution that placed quantum randomness into a new framework for nature's laws should have so dismissed the contingent form of event-chanciness in the macroworld as beyond the pale of scientific interest for being merely historical. 'What is life?' is surely, as Schrödinger held, a question to be answered in the domain of nature's laws. But 'what is life?' is every bit as much a problem in history.

Buckminster Fuller, a modern prophet, often said that 'unity is plural and, at minimum, is two'. Nature's laws and history's contingency must work as equal partners in our quest to answer 'what is life?' For an ancient prophet once stated (Amos 3:3): 'Can two walk together, except they be agreed.'

REFERENCES

Alvarez, L. W., Alvarez, W., Asaro, F. & Michel, H. V. (1980). Extraterrestrial cause for the Cretaceous–Tertiary extinction. *Science* **208**, 1095–1108.

Bowring, S. A., Grotzinger, J. P., Isachsen, C. E., Knoll, A. H., Pelechaty, S. M. & Kolosov, P. (1993). Calibrating rates of early Cambrian evolution. *Science* **261**, 1293–98.

Crow, J. F. (1992). Erwin Schrödinger and the Hornless Cattle Problem. *Genetics* **130**, 237–9.

Darwin, Charles. (1859). *On the Origin of Species*. London: John Murray.

Davies, G. L. H. (1993). Bangs replace whimpers. *Nature* **365**, 115.

Dawkins, R. (1976). *The Selfish Gene*. New York: Oxford University Press.

Dobzhansky, T. (1937). *Genetics and the Origin of Species*. New York: Columbia University Press.

Eldredge, N. & Gould, S. J. (1972). Punctuated equilibria: An alternative to phyletic gradualism. In *Models in Paleobiology*, ed. T. J. M. Schopf, pp. 82–115. San Francisco: Freeman, Cooper & Co.

Gould, S. J. (1982). Introduction, Geneticists and Naturalists. In *Genetics and the Origin of Species*, ed. T. Dobzhansky, pp. xvii–xxxix. New York: Columbia University Press.

Gould, S. J. (1989). *Wonderful Life*. New York: W. W. Norton & Co.

Gould, S. J. & Eldredge, N. (1993). Punctuated equilibrium comes of age. *Nature* **366**, 223–7.
Judson, H. F. (1979). *The Eighth Day of Creation*. New York: Simon and Schuster.
Krogh, T. E., Kamo, S. L., Sharpton, V. L., Marin, L. E. & Hildebrand, A. R. (1993). U-Pb ages of single shocked zircons linking distal K/T ejecta to the Chicxulub crater. *Nature* **366**, 731–4.
Lloyd, E. A. & Gould, S. J. (1993). Species selection on variability. *Proceedings of the National Academy of Sciences USA* **90**, 595–9.
Schrödinger, E. (1944). *What is Life?* Cambridge: Cambridge University Press.
Smocovitis, V. B. (1992). Unifying biology: the evolutionary synthesis and evolutionary biology. *Journal of the History of Biology* **26**, 1–65.
Stanley, S. M. (1975). A theory of evolution above the species level. *Proceedings of the National Academy of Sciences USA* **72**, 646–50.
Stebbins, G. L. & Ayala, F. J. (1981). Is a new evolutionary synthesis necessary? *Science* **216**, 380–7.
Vrba, E. S. & Gould, S. J. (1986). The hierarchical expansion of sorting and selection: Sorting and selection cannot be equated. *Paleobiology* **12**, 217–28.
Whittington, H. B. (1985). *The Burgess Shale*. New Haven, CT: Yale University Press.
Williams, G. C. (1992). *Natural Selection: Domains, Levels, and Challenges*. New York: Oxford University Press.

4

The evolution of human inventiveness

JARED DIAMOND
Department of Physiology, University of California Medical School, Los Angeles, California

How did we humans come to be so different from other animals? That question could not have even been posed until Darwin showed that our differences from animals had evolved. We were not created different from animals. Instead, we had come over time to be different from them.

Until recently, the question how that had happened belonged to the exclusive province of palaeontology and comparative anatomy. Now, insights are flooding in from many other fields, such as molecular biology, linguistics, cognitive psychology, and even art history. As a result, the problem of the evolution of human inventiveness looks as if it may at last be becoming soluble. It is surely among the most challenging questions in biology today.

Despite Darwin, all of us still lump clams, cockroaches, and cuckoos together under an umbrella concept that we term 'animals', and that we contrast with us humans – as if clams, cockroaches, and cuckoos somehow had more in common with each other than they do with us. We thrust even chimpanzees down into that abyss of bestiality, while we stand uniquely on high.

All our unique features are ultimately expressions of our unique inventiveness. Just think of some of the unique forms that our inventiveness takes:

- Unlike any animal, we communicate with each other by means of spoken language and written books.
- We thereby know about things that happened in remote places and at remote past times, such as Schrödinger's 1943 lectures. What animal species possesses any knowledge of what some other individual of its own species on another continent was thinking fifty years ago?
- We depend completely on tools and machines for our living.

- We make and enjoy art.
- We also apply our inventiveness to devising means of genocide, abusing addictive drugs, taking delight in torturing each other, and exterminating other species by the thousands.

No animals species does any of those things. As a result, the laws of Ireland and all other countries insist that, legally and morally, humans are not animals.

Not only are we unique at present: palaeontology teaches us that we are also unique in the history of life on earth. If our differences from animals were just ones of degree, the fossil record might have shown us trilobites wielding compound stone tools in the Palaeozoic era, dinosaurs experimenting with battery-operated rat traps just before the Cretaceous/Tertiary boundary, and baboons developing finger painting in the Miocene era. But all those feats of technology had to wait for *Homo sapiens*.

Palaeontology refutes our historic assumption that intelligence is valuable. Instead, Earth's truly successful animal species, such as beetles and rats, found better routes to their current dominance and wasted little energy on expensive brain tissue. We appear to be unique not only on Earth but also in nearby areas of our Galaxy, since astronomers listening for signs of extraterrestrial intelligence hear nothing but a deafening silence from space.

Despite all that evidence of our uniqueness, it is simultaneously obvious that we are not at all unique. Not only are we indeed animals, but it is even clear what particular sort of animal we are. We are one of the African Great Apes. We have the same anatomical parts as apes, and we have the same or almost the same proteins. Among the proteins sequenced to date in African apes as well as in humans – five haemoglobin chains, myoglobin, cytochrome C, carbonic anhydrase, and fibrinopeptides A and B – most exhibit not even a single amino acid difference between species, and the total number of amino acid changes is only five in 1271 amino acid residues sequenced.* To convince yourself of our kinship to apes, imagine taking some Trinity College students and fellows, putting them into a cage at London Zoo, taking off their clothes, forbidding them to speak to each other, and barring them from visits by a barber for several years. It would

* References for this and other statements will be found in my two earlier explorations of human evolution: *The Rise and Fall of the Third Chimpanzee* (London: Vintage, 1992), and The evolution of human creativity, in *Creative Evolution*, eds. J. Campbell and J. W. Schopf (London: Jones & Bartlett, 1994).

then be obvious that they, and we, are an upright ape with little hair.

From both the fossil and the molecular evidence, we now realize that our ancestors diverged from the ancestors of living African apes only around seven million years ago. That is a mere eye-blink on an evolutionary time scale, much less than 1% of the history of life on earth. As a result, today we are still 98.4% identical in our DNA to the other two species of chimpanzees, the common chimpanzee and the pygmy chimpanzee. Genetically, we are more similar to chimpanzees than Willow Warblers and Chiffchaffs, the two most confusingly similar species of Irish birds, are to each other. If Trinity College had hired an unprejudiced zoologist from Outer Space to classify species, that visitor would have classified us as just a third species of chimpanzee.

Actually, it overstates our distinctiveness to say that we differ by 1.6% from the other chimpanzees, because our unique attributes depend on far less than a 1.6% difference in DNA. Remember that 90% of our DNA is non-coding junk. Remember also that most of the differences between us and chimpanzees in coding DNA have trivial or no consequences for our behaviour, such as the difference of one out of 153 amino acid residues between chimpanzee and human myoglobin. In addition, as we shall see, most of the coding changes in DNA appear to have been completed long before the interesting behavioural differences between humans and chimps even began to emerge. Hence only a tiny fraction of 0.16% of our DNA is likely to explain why we are now discussing evolution in the language of James Joyce's *Ulysses*, instead of foraging speechlessly in the jungle like other chimpanzees.

Which were the few genes that account for that behavioural difference? How did those few genes produce such an enormous difference in behaviour? That is the most fascinating problem of modern biology.

Anybody's first answer is likely to be: the genes responsible for our large brain, the seat of intelligence and inventiveness. Our brain is about four times larger than the chimpanzee brain, and much larger in relation to our body size than the brain of any other animal species. I grant that other attributes beside our large brain were also necessary. Some of them may have provided the initial stimulus to our evolutionary increase in brain size (such as our pelvis modified for walking upright, with subsequent freeing of our hands for other uses). Still other distinctive human characteristics were required to work in concert with our large brains for us to function.

Foremost among those other attributes are our bizarre features of sexual biology (such as menopause, concealed ovulation, and a pair-bonding rare among mammals), which were required for successful rearing of our helpless infants. Yet there is still no disputing the correctness of everybody's first guess, that a large brain was a prerequisite for the evolution of our unique inventiveness.

What is much less appreciated is the fact that our large brain was a necessary, but not a sufficient, condition. That paradox becomes obvious when one compares the time scales for the expansion of brain size and for the appearance of artifacts indicating inventiveness in the human fossil record.

As is well known, evidence from fossil hominids demonstrates that our ancestors had achieved an upright posture around 4 million years ago, that the evolutionary increase in our brain size began by about 2 million years ago, that we had reached the so-called *Homo erectus* grade by about 1.7 million years ago, and that we had reached the archaic *Homo sapiens* grade 0.5 million years ago. The earliest reported anatomically modern *Homo sapiens* – people with skeletons like ours today – lived about 100 000 years ago in southern Africa. At that time, Europe was still occupied by the Neanderthals, who differed significantly in skeletal anatomy and in musculature but whose brain sizes were even slightly larger than those of us moderns.

Thus, our evolutionary increase in brain size began around two million years ago and was essentially complete by about 100 000 years ago. Does archaeological evidence of human inventiveness increase in parallel with that increase in brain size? Such archaeological evidence eventually becomes abundant and includes rock painting, portable art, jewellery, musical instruments, compound tools, intentional burial of the dead, complex weapons such as bows and arrows, complex dwellings, and sewn clothing. If these hallmarks of our inventiveness gradually emerged as our brain size was increasing, then we would have a simple explanation for human inventiveness: it would have been an outcome of our large brain.

Surprisingly, the evidence is unequivocal that that straightforward hypothesis is wrong. The tool kits and diets of those anatomically modern Africans of 100 000 years ago are well documented at South African cave sites. It is clear that they were continuing to make crude stone tools, not at all advanced over those of Neanderthals. Despite their large brains, they were ineffective hunters living at low population densities. Bones of prey mammals represented at their sites consist only of easy-to-hunt animals such as docile antelope, or very young or very old individuals. Dangerous prey species, like rhinoceroses and pigs and elephants, were not yet being hunted. Prey consisted of animals that could safely be killed at close quarters with a hand-held spear, because the spear thrower and the bow and arrow had not yet been invented. Those anatomically modern Africans were taking very few birds or fish as prey, because nets and fish-hooks had not yet been invented. Those large brains were still producing absolutely nothing in the way of surviving art. We would not know it if they had been practising body painting, but they were not producing the art objects that have survived in abundance from somewhat later in the Pleistocene period.

All those hallmarks of inventiveness are likewise absent from sites of the big-brained Neanderthals occupying Europe at the same time. In addition, Neanderthal stone tools show little variation in time and in space. Neanderthal tools from Russia are similar to those from France, while Neanderthal tools from 140 000 years ago are similar to ones from 40 000 years ago. Evidently, the Neanderthals did not exhibit the cultural variation that causes artifacts of *Homo sapiens* today to differ from place to place and from year to year, as a result of human inventiveness.

That evidence for lack of inventiveness is the most astonishing feature of the Neanderthals. For comparison, inventiveness produced such marked cultural differences within the last 10 000 years that archaeologists routinely date sites and group them into real assemblages by their artefacts. As a familiar modern example, styles of computers and cars change so rapidly through invention that they can often be dated to the nearest year. When my computer-literate six-year-old twin sons eventually discover, hidden away in my desk, the slide rule that their computer-illiterate father was using until recently to do his calculations, they will wonder in which phase of the Middle Palaeolithic period I was born.

The only features qualitatively distinguishing human behaviour of 100 000 years ago from the behaviour of animals were the widespread use of those crude stone tools, plus the use of fire. (Chimpanzees also use stone tools, but much less frequently.) At that time we were not even especially successful animals. An Intelligent Extraterrestrial who descended to Earth then would not have singled us out for special mention as creatures with distinctive behaviours, on the verge of taking over the world. Instead, the Extraterrestrial would have singled out beavers, bowerbirds, termites, and army ants. We would have rated no more than passing mention as slightly glorified apes.

What was our large brain doing at that time, when it was still unable to produce archaeological evidence of inventiveness? A flip but, I believe, basically correct answer is that our brain, four times larger than that of chimpanzees, was performing tasks qualitatively similar to a chimpanzee's but was four times smarter. We now know from field studies that chimpanzees make and use tools of a variety of materials (stone, wood, grass); we were making better tools. Chimpanzees and monkeys solve problems better than do other animals, but we would have been solving problems still better. For example, African vervet monkeys for which leopards and pythons are leading predators fail to recognize that a python track in the grass indicates proximity of a python, while a cached carcass in a tree indicates proximity of a leopard; we know better. Chimpanzees use their brains to acquire information about dozens of species, especially plant species, that constitute their diverse diet, including plants with leaves of medicinal value and plants fruiting at long distances and long time intervals. We acquire information about a still broader diet with a high diversity of animal species as well as of plant species. Chimpanzees recognize dozens of individual chimpanzees, tolerate or support individuals from their own troop, kill individuals from other troops, and recognize mother/child associations. We recognize fathers as well as mothers, and we distinguish more complex genetic relationships in addition to those of siblings and parent/child. All those abilities of ours constitute quantitative improvements over chimpanzees and probably drove the evolution of our large brains. But they still do not constitute modern inventiveness or make us qualitatively unique.

In short, by 100 000 years ago many or most humans had brains of modern sizes, and some humans had nearly modern skeletal anatomy. Genetically, those people of 100 000 years ago may have been 99.99% identical to humans today. Despite that close similarity in brain size and other skeletal features, some essential ingredient was still missing. What was that missing ingredient?

That is the biggest unsolved puzzle in human evolution: mostly modern skeletons, and modern-size brains, were not sufficient to produce modern inventiveness.

Now let us jump to western Europe and the period beginning around 38 000 years ago, the time when the first anatomically modern *Homo sapiens* (termed the Cro-Magnons) appeared in western Europe. Beginning then and over the next few tens of thousands of years, the archaeological hallmarks of modern inventiveness appear in western Europe.

Among those hallmarks are the first preserved musical instruments, rock paintings, statuettes and other portable art, clay figurines, and jewellery. The first unequivocal evidence for intentional burial of the dead appears, suggesting the emergence of religion. Tools are no longer the former crude, one-piece, apparently multi-purpose stone tools with no recognizable single function, but instead stone and bone tools with such specialized shapes that their function is still obvious today (functions like needles, fish-hooks, and awls). Compound tools assembled from several pieces appear, such as harpoons, axe-heads hafted onto handles, spears set on spear-throwers, and bows and arrows. Rope is developed and used to make snares and nets; hence fish and birds can now be captured efficiently and appear in abundance in camp remains. Watercraft are invented, as evidenced by the colonization at least 40 000 years ago of Australia and New Guinea, separated by wide and permanent water barriers from the Asian continental shelf. Sewn clothing is depicted in art and evidenced by needles, and makes it possible at last for humans to colonize the Arctic. Archaeological sites include remains of elaborate houses, with paved floors, fireplaces, and post-holes and lit by lamps. An aesthetic sense and desire for luxuries are evidenced by long-distance transport of precious objects such as seashells and superior stone for hundreds of miles across Europe. In contrast, Neanderthal stone tools are made from sources available within a few miles of the site. The most spectacular products of Cro-Magnon inventiveness are those Sistine Chapels of Upper Palaeolithic art, Lascaux Cave and Altamira Cave. A sinister advance in human behaviour is the extinction of 90% of Australia's and New Guinea's large animal species following human colonization, and the extinction of several large mammal species of Europe and Africa. These species had survived at least 20 previous cycles of Pleistocene climate fluctuations, and the only plausible explanation for their disappearance is human arrival (in Australia and New Guinea) or the marked improvement in human hunting skills (in Europe and Africa).

The most significant new trait that appeared in western Europe 38 000 years ago was inventiveness itself. Neanderthal tool types cannot be classified into styles diagnostic of time and place. In contrast, Cro-Magnon tools and art and other cultural products vary so markedly from millennium to millennium and from region to region that archaeologists can use them as indicators of a site's age and affinities. Students of introductory anthropology have to memorize names of Upper Palaeolithic culture horizons, such as the Aurignacian, Gravettian, Solutrean, Magdalenian, and so on. Those names testify to the rapid temporal changes that human inventiveness was at last producing in human cultural products.

We tend to think of the Cro-Magnons as 'cavemen', a word to which our first association is 'primitive'. That association is misleading. We know from technologically 'primitive' people in the modern world, such as New Guinea highlanders dependent until recently on Stone Age technology, that they are fully modern humans in biology and intellect, and that there are simple environmental reasons for their continued use of stone tools. By the same token, I would guess that those Cro-Magnons of 38 000 years ago were also fully modern humans. If they could have been brought into the present by a time machine and sent to Trinity College for an education, they would have learned how to pilot a jet plane or how to become a molecular biologist, just as New Guineans recently out of the Stone Age are doing today. The Cro-Magnons merely had not yet accumulated by 38 000 years ago all the inventions required for airplane technology.

Thus, in Europe there was a sudden Great Leap Forward in human behaviour. In come the Cro-Magnons with all those new behaviours. Within not more than a few thousand years, the Neanderthals who had been occupying Europe for over 100 000 years are gone. Murderers have been convicted on the grounds of less compelling circumstantial evidence. Somehow, the Cro-Magnons undoubtedly caused the disappearance of the Neanderthals, whether by killing or displacing or infecting them.

The cultural revolution that I term the Great Leap Forward looks abrupt in Europe because it was brought in by newly arriving people. The actual Great Leap Forward undoubtedly began outside of Europe and took many thousands of years. Recall that anatomically modern *Homo sapiens* already existed in Africa and the Near East around 100 000 years ago, and coexisted with Neanderthals in the Near East for a long time without being able to exterminate the Neanderthals. Probably all those distinctions of the Cro-Magnons developed between 100 000 and 38 000 years ago in Africa, or in the Near East, or Asia, or somewhere else, and were then imported into Europe. But even the period from 100 000 to 38 000 years ago is a tiny fraction of the seven million years since our ancestors diverged from the ancestors of chimpanzees. What was that last 0.01% of our genes that changed during that brief time, and that caused the Great Leap Forward?

There is only one hypothesis that seems plausible to me: the genes responsible for the perfection of spoken language. Many animal species have systems of vocal communication, but none is remotely as sophisticated and expressive as human language. It is striking that chimpanzees and gorillas have been *taught* to express themselves with computer languages or sign languages comprising hundreds of symbols. That symbolic repertoire is nearly as large as the 600 words that constitute the daily working vocabulary of the average American and Englishman. Pygmy chimpanzees have been taught to *understand* instructions in spoken English, conveyed in a normal tone of voice and sentence structure. Thus, apes clearly possess some of the capacity required for language.

Nevertheless, chimpanzees and gorillas do not and cannot speak. Even an infant chimpanzee brought up in the house of a husband/wife psychologist couple, together with the couple's human child of the same age, was never able to learn to utter more than a couple of different vowels and consonants. That limitation stems from the structure of the ape larynx and vocal tract. To satisfy yourself how that limits expressiveness and inventiveness, try seeing how many different words you could speak if you could only pronounce the vowels *a* and *u*, and the consonants *c* and *p*. If you wanted to say 'Trinity College is a fine place to work', all you could manage would be 'Capupa Cappap up a cap capcupap'. Your attempt to say 'Trinity College is a bad place to sneeze' would result in identical sounds.

Without language, we cannot communicate a complex plan, nor think out the complex plan in the first place, nor brainstorm about how to design a better tool, nor discuss a beautiful painting. But our vocal tract is like a fine Swiss watch, with dozens of tiny muscles, bones, nerves, and pieces of cartilage working together in precisely coordinated ways. Thus, given an ancestral human that already had four times the brain capacity of a chimpanzee, and given the already impressive linguistic capabilities of chimpanzees, a series of small changes in the structure of the vocal tract, which let us pronounce dozens of distinct sounds instead of just a few, may have been the trigger for complex language and hence for the Great Leap Forward. Those small changes may have been the last missing prerequisite for the evolution of human inventiveness.

With language, we can invent. The essence of human language is inventiveness: every sentence is a new invention, produced by combining familiar elements. For that reason, it is inconceivable to me that those non-inventive humans of 100 000 years ago could have had language as we know it today. I cannot avoid the conclusion that the development of human inventiveness was linked to the perfection of human language.

If we accept that reasoning, can we recognize any intermediate stages in that development of modern human language from its precursors of animal vocal communication systems? At first, it seems as if an unbridgeable gulf separates the barking of dogs from the language of James Joyce's *Ulysses*. In fact, studies over the last two decades have identified at least three intermediate stages within that gulf.

One early stage is the 'language' of wild vervet monkeys, a common monkey species of East Africa. When you listen to vervets, it at first appears that they are uttering undifferentiated grunts. But if you listen carefully, you may be able to distinguish differences among the grunts. Experiments with tape recording and playback of wild vervet vocalizations revealed that the vervets have at least ten different grunts, including separate 'words' for their three major predators (leopards, snakes, and eagles), separate words for minor predators (baboons, other predatory mammals, and strange humans), and separate words for various social categories of vervets (dominant monkey, subordinate monkey, rival monkey).

There is no reason to believe that vervets are unique in possessing such a natural language. Instead, the recognition of vervet language was long delayed because vervets are much better tuned to the distinctions among their grunts than we are. To decode their grunts required tape recorder playback experiments, facilitated by the open habitat and small territories in which vervets live. It seems likely that wild chimpanzees and gorillas will also prove to have natural languages, but those have not yet been identified because of the severe logistical problems presented by their denser habitat and much larger territories.

While vervet language includes different sounds for different meanings, it nevertheless lacks the essential structure of modern human language: the latter's modular hierarchical design. By that, I mean that we combine units of a few dozen vowels and consonants into higher units of a hundred or so different syllables, combined in turn into higher units of thousands of different words, organized in turn into phrases, and organized again to produce an infinite number of possible sentences. That hierarchical combination is carried out according to grammatical rules for constructing and combining words. So far, no such hierarchical design has been detected in vervet language: they appear to communicate by uncombined single sound units.

The natural language of vervet monkeys thus illustrates a likely early stage in the development of human speech. We repeat that stage ontogenetically in the course of speech acquisition by human infants, who begin with single 'words' uttered in isolation. Can we now go to the other end of the postulated animal/human language continuum and recognize, as other intermediate stages, some simple human languages less complex than normal human speech? Do any primitive human languages still exist in the world today?

Nineteenth-century explorers repeatedly made such claims. They returned from remote areas of the world, reporting that they had discovered primitive tribes with primitive technology, people so primitive that they communicated only with monosyllabic grunts like 'ugh'.

All those stories proved to be false: all existing normal human languages are fully modern and expressive. Technologically primitive peoples do not have primitive languages. In fact, the languages of the New Guinea highlanders with whom I have been working in my field studies of bird evolution, and who were still dependent on stone tools as recently as the 1970s, are without exception much more complex grammatically than English and Chinese, languages that we associate with civilization. Alternatively, we can examine the most ancient languages that have come down to us through preserved writing, the earliest written Sumerian of 3100 BC and Egyptian of 3000 BC. Those first written languages, too, were already typical modern languages in their complexity. Thus, it appears that human language had already reached its modern complexity long before 3100 BC, and that there are no remaining primitive human languages to suggest how vervet-like languages evolved into the language of *Ulysses*.

In fact, there actually are spoken today some simple human languages, much more complex than vervet language, but much less complex than normal human languages. Those simple languages are ones that have been spontaneously invented innumerable times in human history, whenever people who did not share a common language were thrown together, such as traders and the native peoples with whom they were trading, or else plantation overseers and mixed groups of plantation workers of different origins. Within a few years in each case, the trading parties or the workers and overseers who were thrown together evolved a rudimentary language for communicating with each other, termed a pidgin language. Pidgins just consist of strings of words with little grammar or phrase construction, and composed mainly just of nouns, verbs, and adjectives. This stage, too, corresponds to one that we traverse ontogenetically, as young children progress from vervet-like single utterances to word strings. While pidgin languages are used by trading partners or by workers and overseers to communicate with each other, each group continues to use its own normal complex language for communication within the group.

Pidgin languages are adequate for the circumscribed range of meanings that they serve to communicate. However, the children of pidgin-speaking parents face a big problem in communicating with each other, because pidgins are so rudimentary and inexpressive. In such situations the first generation of children of pidgin-speaking parents spontaneously develops a pidgin into a more complex language termed a creole, which becomes spontaneously stabilized within one generation. I emphasize that the evolution of a pidgin into a creole is unplanned and spontaneous. It is not the case that the children sit down, acknowledge that their parents' language is inadequate, and then decide among themselves which child will invent pronouns while other children are working out the pluperfect conditional tense.

Creoles are fully expressive languages with newly invented grammars, and with the modular hierarchical organization characteristic of normal human languages. As an example, consider the following sentence that I encountered in a creole language:

Kam insait long stua bilong mipela – stua bilong salim olgeta samting – mipela i-can helpim yu long kisim wanem samting yu likem, bigpela na liklik, long gutpela prais.

That sentence was one that I read in a supermarket advertisement in Port Moresby, the capital of Papua New Guinea, in the creole language termed Neo-Melanesian. The English translation of that advertisement is as follows:

Come into our store – a store for selling everything – we can help you get whatever you want, big or small, at a good price.

Comparison of this translation with the original will show that the creole text has a full modular hierarchical structure and includes such sophisticated grammatical elements as conjunctions, pronouns, relative clauses, auxiliary verbs, and imperatives.

All over the world, similar creoles have arisen repeatedly from pidgins with the most varied vocabularies and speakers. Pidgin speakers have variously included Africans, Chinese, Europeans, and Pacific islanders, while the language providing much of the vocabulary has variously been Arabic, English, French, Portuguese, and German. Despite those enormous differences among creoles and their origins, resulting in entirely different *vocabularies* for creoles of different origins, the *grammars* of the resulting creoles are all quite similar, both in what they lack and in what they possess. Compared to many normal languages, creoles lack verb conjugations for person and tense, noun declensions for case and number, and most prepositions. However, creoles share with most normal languages their possession of relative clauses, singular and plural first-person, second-person, and third-person pronouns, and particles or auxiliary verbs expressing negation, anterior tense, conditional mood, and continuing action, placed in approximately the same sequence.

Thus, creoles share striking similarities in their grammar, despite their independent origins and differing vocabularies. They evidently spring out of some genetic hard-wiring of a universal grammar inside our brains. As most of us are growing up as children, we hear a normal complex language spoken around us, and we learn that language, which overrides the genetically hard-wired universal creole grammar. Only those children who grow up in an environment where no complex language is spoken have to fall back on that hard-wired grammar.

Thus, vervet language, pidgins, and creoles represent three stepping-stones that may exemplify how complex modern human languages evolved from animal precursors. I would guess that the delay between about 100 000 and 40 000 years ago, between the first appearance of full-sized brains and anatomically modern skeletons and the later first appearance of modern human inventiveness, was mostly due to the time required to perfect modern hierarchical language. I would guess that, if we could devise a time machine with a tape recorder to place in the camps of *Homo erectus* and Neanderthals, we would find that they spoke in pidgins, with few distinct sounds and little grammar to structure their word strings. Between 100 000 and 40 000 years ago, we may have been perfecting the anatomy of our vocal tract, so as to be able to enunciate distinctly dozens of vowels and consonants. We may also have been perfecting the organization of those vowels and consonants into syllables and words, and organizing those words in phrases and sentences unlike the word strings of *Homo erectus* and Neanderthal pidgins.

Finally, we may have been evolving a universal grammar, and getting it genetically hard-wired inside us.

Laboratory scientists tend to dismiss historical sciences, such as evolutionary biology, as soft or speculative. Yes, it is more difficult to arrive at knowledge in fields where the methodology of the controlled and replicated laboratory experiment that manipulates a well-designed test system cannot be applied. Nevertheless, the historical sciences have developed their own successful methodologies. In the next 50 years, what techniques may help us understand the evolution of human inventiveness?

Some progress will undoubtedly come from dramatic new advances. For example, the human genome is now being sequenced, and DNA has been successfully extracted from plants and animals thousands or even millions of years old. The recent discovery of a 5000-year-old Copper-Age mummy from the Alps permits us to dream that a 30 000-year-old mummy could also be discovered. Perhaps current efforts to extract DNA from dried blood or tissue will be successful. In that case, we might actually be able to compare the DNA of modern humans, vanished human ancestors, and chimpanzees.

But we are also likely to learn much from extensions of methods already in hand now. I mentioned the discovery of a natural language of vervet monkeys, and the technical problems confronting attempts to study natural languages of wild chimpanzees and gorillas. It seems only a matter of time before someone will tackle that problem of ape natural language. A second development is that methods for studying ape cognition by allowing apes to communicate via computers have advanced rapidly in the past decade. A third promising area is that linguists are now attempting to discern relationships among human languages that diverged more than 10 000 years ago, and perhaps to reconstruct human proto-languages of the remote past. Finally, it has become possible only within the past few years to date Upper Palaeolithic rock paintings by carbon-14 dating of the paint materials themselves. These results are just beginning to yield insights into the sequence of development of human art techniques, a window into human creativity.

In short, what seems to me perhaps the most challenging problem for biology today is posed by a historical decoupling. In our evolutionary history, changes in human brain size and human skeletal anatomy were decoupled from changes in human inventiveness, as gauged by the artifacts that our ancestors left behind. Our increase in brain size, and most of the development of our modern skeletons, were virtually complete tens of thousands of years before most types of evidence for human inventiveness even begin to appear. Those recent types of evidence include art, rapid cultural changes in time and space, burial of the dead, and long-distance trade.

The entire genetic difference between us and the other two species of chimpanzees comprises only 1.6% of our genome. The total difference in coding DNA is probably only about one-tenth of that, and the coding changes left to be completed after 100 000 years ago were far less than even that. The best guess that I can make, about those final changes responsible for our Great Leap Forward in behaviour, involves the perfection of modern language. If so, those final changes were the main reason why *we* are now sitting in Trinity College, using the language of James Joyce to discuss primate evolution, while our closest relatives the chimpanzees are at the same time eating termites in the jungle or held captive by us in zoos.

5

Development: is the egg computable or could we generate an angel or a dinosaur?

LEWIS WOLPERT
Department of Anatomy and Developmental Biology, University College London

In calling the structure of the chromosome fibres a code-script we mean that the all-penetrating mind, once conceived by Laplace, to which every causal connection lay immediately open, could tell from their structure whether the egg would develop, under suitable conditions, into a black cock or into a speckled hen, into a fly or a maize plant, a rhododendron, a beetle, a mouse or a woman.

What we wish to illustrate is simply that with the molecular picture of the gene it is no longer inconceivable that the miniature cell should precisely correspond with a highly complicated and specified plan of development and should somehow contain the means to put it into operation.

(E. Schrödinger, 1944)

These quotations from Schrödinger were very perceptive, and raised two key questions. The first is whether the development of the egg is computable, and I will suggest that the answer is no, but that it will be possible to simulate some aspects of development. As regards the second question, how genes control development, Schrödinger could not have known that genes exert their influence through controlling which proteins are made, and in this way they control cell behaviour and development.

In posing these questions, Schrödinger was giving recognition to the fundamental importance of development. Development is at the core of multicellular biology. It is the link between genetics and morphology. Indeed much of the genetic information in our cells is required to direct development. Evolution can be thought of in terms of altering the programme of development so that structures are modified and new ones form. It is only the genes that change in evolution, so understanding how genes control development is fundamental to understanding the evolution of animals and plants. When we know this then we could consider whether we could generate either an angel or a dinosaur.

It is interesting to briefly compare embryology of fifty years ago to the present. At that time Needham's book *Biochemistry and Morphogenesis* (1942) had just been published. It was largely concerned with what now seems to be irrelevant biochemistry, and the search for inducing substances, and signalling molecules. Today while genetics and molecular biology have transformed the field we have to recognize that in only a couple of cases can we point with any confidence to signal molecules: the *bride of sevenless* in the insect eye; TGF-β-like molecules in the development of the insect gut (Lawrence, 1992); and some molecules in the development of the vulva in nematodes. In vertebrates we do not have a single well established case of an inducer molecule or morphogen; lots of hopeful candidates but nothing conclusive. By contrast, Huxley and De Beers' earlier book *The Elements of Experimental Embryology* (1934), which contained virtually no biochemistry, is much more relevant to current thinking with its emphasis on gradients and interactions.

The key to development is the cell – the true 'miracle' of evolution. One can make a strong case that given the eukaryotic cell, its elaboration to generate multicellular animals and plants was by comparison easy. For example, the cell cycle and cell division can be thought of as a developmental programme. Development is merely the modification of cell behaviour and in a sense one can regard the cell as more complex than the embryo: more complex because the interactions between the parts of the embryo are very much simpler than the interactions between the components in the cell. One should think of all cell interactions in the embryo as selective rather than instructive. The interactions merely select from one of the possible states which the cell can adopt; usually these are few, two or three, though on rare occasions more numerous. The interactions provide the cells with rather low level information. The complexity of development lies in the internal programme of the cells.

The evolution of development is an important topic in its own right: what are the selective pressures on development and how does novelty arise? I have argued that embryos may be evolutionarily privileged. That is, since they only have to develop reliably and provided this occurs, they may be able to explore developmental possibilities without negative selection (Wolpert, 1990).

Since proteins essentially determine cell behaviour, development can be thought of as controlling which proteins are made where, and so controlling the activity of the genes coding for them. How many genes are involved in controlling development as distinct from providing the household functions of the cell? Of course the answer is not known but one can make some informed guesses. An estimate of the number of genes in *Escherichia coli* is 4000, yeast 7000, and in the nematode 15 000 (Chothia, 1992). It is not unreasonable to think that of the 60 000 genes in humans some 30 000 may be involved in development. By contrast, analysis of early insect development suggests that only about 100 genes are involved in controlling pattern during early development. And in the nematode about 50 genes are known that control vulval development. While these are much smaller numbers, if one thinks of say 100 genes per structure then 50 different structures in *Drosophila* would require 5000 genes. Another way of looking at gene number is in relation to the number of cell types. In humans, there are about 250 different cell types and if each is characterized by 10 different proteins, and each protein requires 10 genes for its specification, both being quite modest numbers, then we already arrive at 25 000 genes for development. Structures like the brain might require a much larger number. It is also unlikely that there is much overlap between the genes for different 'organs' for this would result in a lack of flexibility in evolution, there would be too much pleiotropy.

Tens of thousands is a large number of genes whose action needs to be understood. Understanding is made even more difficult by cases of apparent redundancy. That is, it is possible to knock out certain genes in mice without there being any obvious phenotype. So what is the function of such an apparently redundant gene? I have argued that all redundancy is illusory and merely reflects the failure to provide the correct test for the altered phenotype (Wolpert, 1992). Even a 5% disability requires the examination of 20 000 animals. It will be very difficult to work out the true function of such genes.

To what extent can we expect general principles to emerge over the next fifty years or are we simply faced with a long period of collecting detail? At present we can put forward a list of rule-like ideas and we feel that we basically understand the fundamental principles of development, and it is striking how few concepts are required. A central assumption is that the state of a cell is determined by which genes are on and thus which of their proteins are present in the cell. While protein and mRNA degradation may be important, as well as translational control, it is a good starting point. Probably one of the key integrating structures in development is the promoter and enhancer chromosomal region. This upstream control region has undergone major evolutionary changes and may serve to integrate many aspects of cell behaviour. For example, it seems that spatial localization of gene expression in insect development is the result of different factors binding to an enhancer region and providing a threshold response to external signals (Lawrence, 1992).

Development is largely about cells becoming different in an ordered manner. One can think of the earliest multicellular organisms solving this in two ways: one is by asymmetric cell division, the other is by cell interactions (Wolpert, 1990). These are the only ways in which differences arise and it remains a puzzle as to why animals use one rather than another. Many animals develop along Cartesian axes, pattern being specified independently along each. One way of making patterns involves assigning cells positional information as in a coordinate system and the cells then interpreting these values in a variety of ways. This has the important implication that there is no relationship between the early pattern and the observed one. Another common feature seems to be the generation of periodic structures like segments, vertebrae, feathers and teeth, which are built on a basic ground plan which is modified by positional information. All the interactions are short range – rarely over more than 30 cell diameters – and most patterning occurs locally so that embryos are quickly divided into regions which develop largely independently of one another.

Our best system for understanding development is the fruit fly *Drosophila* (Lawrence, 1992). The two axes, the antero-posterior and dorso-ventral, are initially independent of one another and are specified by maternal gene products which provide gradients of positional information. After fertilization the gradients activate a cascade of zygotic genes and the embryo becomes divided up into a number of regions defined by combination of the activity of different genes. Along the antero-posterior axis, a periodic pattern of gene activity is established – the forerunner of segments. Remarkably each stripe is specified independently by the local combination of proteins. Each segment also acquires a unique identity coded for by the activity of a special set of genes known as the *Hox* genes. Another aspect of fly development which is an excellent model is the ommatidium of the eye. Here 8 cells form a photoreceptor complex, each of the 8 cells having a unique identity. Some of the genes and signals have been identified. Unlike a patterning mechanism based on positional information, it seems that there is a sequence of cell interactions so that each of the 8 cells is specified in the correct place. The interactions thus only involve signalling from one cell to its neighbours. Slightly longer range signalling is involved in spacing individual ommatidia.

Spatial organization and generating differences dominate early development and in general precede and specify morphogenesis (or change in form) and cell differentiation. Morphogenesis is about cellular forces changing the shape and relationships between cells while differentiation leads to the production of molecules that characterize different cell types.

Change in form is a problem of linking gene action to mechanics. While there is preliminary understanding of the cellular forces involved in gastrulation in amphibia, insects and sea urchins we have little knowledge of the intracellular machinery involved, how the movements are coordinated, and how they are initiated at the right time and place. We need to know how genes can control cellular forces. One obvious mechanism is by controlling the spatial pattern of expression of cell adhesion molecules.

Of considerable importance is the extent to which developmental mechanisms have been conserved. This is particularly well illustrated with respect to the role of homeobox genes providing positional identities for cells along the antero-posterior axis. This relates to a general principle, namely that patterning often occurs in two main steps – assigning positional identities and then the cells interpreting these in a variety of ways. Thus the similarity in *Hox* gene expression along the antero-posterior axes of vertebrates and flies is incomparably greater than the structures which later develop. Thus there is convergence towards establishing similar positional values along the axis, even though different mechanisms may be used, and then divergence of later development. There is also probably a high degree of conservation of morphogenetic mechanisms – cell adhesion and cell contractility being used again and again. One need only look at the similarity in gastrulation of insects and sea urchins. But as regards cell differentiation it is not clear what general principles are involved since differentiation is essentially the control of expression of cell specific proteins. One would expect to find the greatest divergence here for there is nothing similar which may be expected to be found in say the differentiation of muscle and red blood cells other than the activation of cell and tissue specific transcription factors.

Currently we are riding on a wave of excitement following the identification of the *Hox* genes, possible signalling molecules, and detailed analysis of some developmental systems like early fly development, the development of the fly eye and the nematode. We have the feeling, perhaps an illusion, that we understand the basic principles controlling development. We can see how cascades of gene action and intracellular signalling can generate pattern. Even with the limb there are quite plausible models involving homeobox genes and growth factors (Wolpert & Tickle, 1993). Against that must be set our ignorance: there is not a single case in vertebrates where a signal molecule has been unequivocally identified; our understanding of cell structure with respect to the establishment of polarity is still primitive; so too is our understanding of morphogenesis at the molecular level. There are plausible models for gastrulation in flies, sea urchins and amphibians but the molecular basis and their genetic control is lacking. We are also particularly weak in our understanding of features like the regulation of size and form. But we think understanding of all these will come with better knowledge of cell biology. It is striking that ciliated protozoa develop complex patterns and obey rules similar to multicellular organisms, yet we have no understanding of the molecular mechanisms (Frankel, 1989). And while the specification of positional identity by *Hox* genes is encouraging the interpretation of this positional information, the downstream targets of the *Hox* genes, are, in general, not known, particularly in relation to morphogenesis: the alteration of just one gene can, in the fly, alter an antenna into a leg.

Is the egg computable? That is, given a total description of a fertilized egg – the total DNA sequence and the location of all proteins and RNAs – could one predict how the embryo would develop? Can we anticipate general theories of development and what will these theories look like? One's judgement will, at this stage, reflect one's view of the developing embryo. Is it best treated as a dynamical system or a finite state machine? If it is treated as a dynamical system then it is possible that theorems from that discipline may be relevant, theorems about attractors and limit cycles (Kelso, Ding & Schöner, 1992). Such dynamical systems are based on non-linear dynamics and analyse non-equilibrium chemical processes in terms of fluctuation and instability and, particularly, self-organization of spatial and temporal patterns. A characteristic feature of all such systems is that they seem to exclude structure from the initial conditions. Yet both cells and embryos are highly structured. Perhaps more important is that they treat these systems as if they were continuous, but cell behaviour and development is largely based on switches. Turning a gene on is a switch that can result in the production of a new protein which can completely alter the behaviour of the cell. It is also notable that as yet the dynamical system theory approach has not so far been fruitful in cell or developmental biology. A potential exception is reaction–diffusion along the lines suggested by Alan Turing. Reaction–diffusion mechanisms provide an attractive model for self-organization of gradients and periodic structures (Murray, 1989) but as yet there is no persuasive evidence that it operates during development.

A counter-example to a dynamical system is the self-assembly of bacteriophage which is hard-wired into the amino-acid sequence of the proteins and shows an obligatory pathway for protein interactions. The same is probably true of the assembly of cellular organelles like ribosomes, actin filaments and collagen. Such self-assembly most likely is involved in cellular differentiation like filament assembly in muscle cells.

If development is to be regarded as a finite state machine, Wolfram's (1984) models of cellular automata may be quite instructive. Instead of models based on differential equations which describe smooth variations of parameters in relation to each other, cellular automata are based on the discrete changes in many similar components. While some cellular automata can be analysed as a discrete dynamical system, there are some for which the only way to determine their development is by simulation: no finite formulae for their general behaviour could be given. Even quite simple rules based on the values of neighbours lead to patterns that are not computable in the sense of predicting the outcome without actually seeing how the system balances.

Is development similar to a non-computable cellular automaton? It seems very likely that in some ways it is. Cell behaviour during development is determined by the cell's current state and the signals from its neighbours. These determine its next state. All these states can best be characterized by which of the genes are on. Nevertheless one must take into account that there may be complex interactions between one state and the next: for example, a new protein produced by turning on a gene may modify and interact with other proteins leading to a cascade of events which together with the initial protein can modify the cell behaviour that leads to the next state. This emphasizes an important difference between development and cellular automata. For instead of there being just a few states whose pattern varies with successive generations, in development new cell states are continuously being generated. There are thus thousands of different cell states as defined by different patterns of gene activity during the development of an embryo. The development of an embryo is much more complex than that of cellular automata and this complexity arises because of the complexity of cells and the large number of different states they can exhibit. It thus seems unlikely that it will be possible to even formally simulate development.

It will nevertheless be important in the future to try and simulate processes involving change in form, like gastrulation. The movements are slow and do not involve an initial component and so the system can be regarded as quasi-static. This may simplify attempts to simulate morphogenetic movements. But simulating even gastrulation in a vertebrate is a major task and simulating some aspect of organogenesis, like that of the brain, is more daunting still.

In fifty years' time will we yet be able to fully determine the initial conditions? We will by then know the complete DNA sequence but we will need to know much more. We will need to know what proteins and maternal messages are stored in the cytoplasm and their spatial distribution. Quite small variations could be significant and might be very hard to detect. We will also need to understand the complex interactions that are involved in intracellular signalling and the role of the myriad kinases and phosphatases. We may be able to neglect metabolism but this is far from clear. But the central point is that any detailed understanding or computation of development will require detailed understanding of cell biology. This is a formidable task for it implies that in computing the embryo it may be necessary to compute the behaviour of all the constituent cells. There may however be a simplification if a level of description of cell behaviour can be found that is adequate to account for development and that does not require each cell's detailed behaviour to be taken into account.

An analogy to some of these problems is protein folding itself, which seems a much simpler problem. Will it be possible in the next 50 years to predict the three-dimensional structure of a protein from its amino acid sequence? The answer is probably yes but not necessarily by working the structure out from, as it were, first principles. Rather the solution will come from homology. Chothia (1992) has pointed out that proteins are derived from about 1000 protein families and as the rules for the folding of each of these are worked out by crystallography, NMR and molecular modelling, the structure of any new protein will probably be predictable.

Similar principles may well hold for predicting how an embryo will develop. The early development of different organisms can be very different. So, even though one could identify homeobox genes involved in axial patterning, it will be very difficult to work out their spatial pattern of expression.

If one did know this pattern, then together with the knowledge of the genes to whose controlling regions their products bind, it might just be possible to make some general predictions of what sort of animal would develop. Like protein folding, homology based on an extensive database could provide the best basis for making such predictions. Thus, while there may be general principles and the same genes and signals used in diverse organisms, the details will be all important and will make predictions about development particularly difficult. In spite of all this it is not unreasonable to think that eventually enough will be known to program a computer and simulate some aspects of development. We will, however, understand much more than we can predict. For example, if a mutation were introduced that altered the structure of a single protein, it is unlikely that it will be possible to predict its consequences.

So what will the next fifty years bring? If we are correct in believing that we understand the basic mechanisms of development, then no new principles may emerge, but rather it will be fifty years of hard slog working out the fine details of cell behaviour during development. This will include a detailed understanding not only of gene action but the biochemistry and biophysics of cells. This detail can, however, be very exciting. Such a prediction is both pessimistic and optimistic: optimistic because it would mean we understand the principles of development, and pessimistic because the future looks a little tedious. The truth almost certainly lies somewhere between these positions and it will be disappointing and surprising if no new mechanisms or ways of integrating the information emerge. Also, powerful new techniques will certainly be invented.

One of the pleasures of being able to simulate development, if it were possible, is the impact it will have on our understanding of evolution. We could, for example, ask what sequence of genetic changes could have lead to, for example, the evolution of limbs or the brain. We could 'play' on the computer to see the effects of altering one gene at a time. We could, in principle, try to devise a genetic program which would generate a dinosaur or an angel. The problem with the angel is to provide both an extra pair of wings as well as an angelic temperament. To get an extra pair of feathered limbs would require considerable ingenuity but if we knew enough about the patterning of the body plan, and the development of wings and feathers, it would be plausible. One would be making use of known genes from birds and mammals. It is most unlikely we would know what neuronal connections to generate to get an angelic temperament, but one could probably devise a selection procedure given enough time. A dinosaur would be even more difficult even if we had the complete DNA. The problem would be to establish the correct initial conditions for dinosaur development. *Jurassic Park* will remain science fiction.

REFERENCES

Chothia, C. (1992). One thousand families for the molecular biologist. *Nature* **357**, 543–544.

Frankel, J. (1989). *Pattern Formation*. New York: Oxford University Press.

Huxley, J. S. & De Beer, G. R. (1934) *The Elements of Experimental Embryology*. Cambridge: Cambridge University Press.

Kelso, J. A. S., Ding, M. & Schöner, G. (1992). Dynamic pattern formation: a primer. In, *Principles of Organization in Organisms*, eds. J. Mittenthal & A. Baskin, pp. 397–439. Reading, MA: Addison Wesley.

Lawrence, P. A. (1992). *The Making of a Fly*. Oxford: Blackwell.

Murray, J. D. (1989). *Mathematical Biology*. New York: Springer.

Needham, J. (1942). *Biochemistry and Morphogenesis*. Cambridge: Cambridge University Press.

Schrödinger, E. (1944). *What is Life?*. Reprinted (1967) with *Mind* and *Matter and Autobiographical Sketches*. Cambridge: Cambridge University Press.

Wolfram, S. (1984). Cellular automata as models of complexity. *Nature* **311**, 419–424.

Wolpert, L. (1990). The evolution of development. *Biological Journal of the Linnaean Society* **39**, 109–124.

Wolpert, L. (1992). Gastrulation and the evolution of development. Development Supplement 7–13.

Wolpert, L. & Tickle, C. (1993). Pattern formation and limb morphogenesis. In *Molecular Basis of Morphogenesis*, ed. M. Bernfield, pp. 207–220. New York: Wiley-Liss.

6

Language and life

JOHN MAYNARD SMITH[1] and EÖRS SZATHMÁRY[2]
[1]*Department of Biology, University of Sussex, Falmer, Brighton*
[2]*Department of Plant Taxonomy and Ecology, Eötvos University, Budapest*

All living organisms can transmit information between generations. The property of heredity – that like begets like – depends on this transmission of information, and in turn heredity ensures that populations will evolve by natural selection. If we ever encounter, elsewhere in the galaxy, living organisms derived from an origin separate from our own, we can be confident that they too will have heredity, and a language whereby hereditary information is transmitted. The need for such a language was central to Schrödinger's argument in *What is Life?*: he referred to it as a 'codescript'. We can make some guesses about the nature of the language. It will be digital, because a message encoded in continuously varying symbols rapidly decays into noise as it is transmitted from individual to individual. It must also be capable of encoding an indefinitely large number of messages. These messages must be copied, or replicated, with a high degree of accuracy. Finally, the messages must have some 'meaning', in the sense of influencing their own chances of survival and replication: otherwise, natural selection will not operate.

In existing organisms, there are two such languages, not one. There is the familiar genetic language based on the replication of nucleic acids, DNA and RNA, and there is the even more familiar language, confined to humans, which we are using now. The former is the basis of biological evolution, and the latter of cultural change. In this essay we discuss the origins of both.

In fact, we will not discuss the origin of nucleic acid replication, although this was a crucial step – perhaps *the* crucial step – in the origin of life. Instead, we discuss the origin of the genetic code. In all existing organisms, there is a division of labour between nucleic acids and proteins. Nucleic acids carry genetic information, which is transmitted through replication.

Proteins determine the phentoype of the organism. The connection between the two is *via* the genetic code, whereby the base sequence of a nucleic acid is translated into the amino acid sequence of a protein. It is in this process of translation that nucleic acids acquire what we have called their meaning: by specifying proteins, they influence their chances of survival – their 'fitness'. The mechanism of translation is at the same time so complex, and so universal, that it is hard to see how it could have originated, or how life could have existed without it.

The second of these problems, the existence of life without the code, which seemed an almost impenetrable mystery ten years ago, is no longer so mysterious. The crucial discovery is that, even in existing organisms, some enzymes are made of RNA, not protein (Zaug & Cech, 1986). This has led to the idea of an 'RNA world', in which the same RNA molecules were both phenotype and genotype, both enzymes and carriers of genetic information. Given this picture, which we accept, it is possible to have life without proteins, and hence without the code. It is also easier to imagine how the code might have originated.

The essential feature of the code is that each triplet of nucleotides – each 'codon' – is assigned to one of 20 amino acids. This assignment is brought about by the attachment of particular amino acids to particular tRNA molecules, each incorporating the relevant codon. This attachment is carried out by specific enzymes, which can be called assignment enzymes. The specificity of the code depends on the specificity of these enzymes. Our problem is to explain how this specificity originated.

Before turning to this question, however, we review briefly what can be deduced from the nature of the existing code. There are some variations: for example, in yeast and most animal mitochondria, the codon AUA codes for methionine instead of isoleucine. A number of such differences are known, and more will probably be discovered. The variability is limited, however, and is consistent with the idea that there was a single ancestral code, and that there have been a number of minor departures from it. The existence of variations does raise one problem. How can the code evolve? If, for example, AUA codes for isoleucine, as it does in the 'universal' code, how could the assignment change? The snag is that there are, typically, AUA codons at many sites in the genome of an organism. Even if it were selectively advantageous to change isoleucine to methionine at one of these sites, it would surely be disadvantageous to make the change at all of them. Possible mechanisms of change are reviewed by Osawa *et al.* (1992). In essence, they suggest that directional mutation pressure, which alters the ratio of adenine–thymine base pairs to guanine–cytosine base pairs, leads to particular codons no longer being used: an unused codon can then be re-assigned.

The important point is that the code can evolve, albeit rarely and with difficulty. During early evolution, when organisms were simpler and had few genes, evolutionary change was probably easier. The significance of this is as follows. As we show in a moment, the code has some adaptive features. In general, evolutionary biologists explain adaptation by natural selection. A code that could not change could not become adaptive in this way. But if, as seems to be the case, the code can evolve, these adaptive features are easier to explain.

The clearest example of an adaptive feature is this: chemically similar amino acids tend to be coded for by similar codons. For example, aspartic and glutamic acids are chemically similar: aspartic acid is coded for by the codons GAU and GAC, and glutamic acid by the codons GAA and GAG. More general analysis confirms that, in this respect, the code is far from random. Why should it be adaptive for similar amino acids to be coded for by similar codons? Two plausible reasons have been suggested. First, if an error is made in protein synthesis, the effect on protein function is likely to be relatively slight. Second, mutations are less likely to be damaging.

A second non-random feature of the code concerns its redundancy. Amino acids may be coded for by 1, 2, 3, 4 or 6 different codons. In general, amino acids that are common in proteins tend to be specified by more codons: for example, leucine and serine (both 6 codons) are commoner in proteins than tryptophan (1 codon). But it would probably be wrong to interpret this as an adaptive feature of the code. It is more likely that it is an unselected consequence of the code being as it is. Thus there will be more mutations to serine and leucine than to tryptophan. If at least some amino acid changes are selectively neutral, the observed association between abundance in protein and redundancy is to be expected. There is also clear evidence that selection has prevented the abundance in protein exactly matching redundancy. For example, the frequencies of the acidic (Asp and Glu) and basic (Arg and Lys) amino acids are approximately equal, as would be expected, since intra-cellular pH is neutral. But on the basis of codon redundancy one would expect the basic amino acids to be twice as frequent.

There remains the question whether there is any chemical reason why particular codons became associated with particular amino acids. The alternative is that the assignments were chemically arbitrary, as assignments of meanings to words in human language are largely arbitrary. On the latter assumption, there may be a reason why the first two nucleotides in the codons for Glu and Asp are the same, but it is pure accident that they are GA, and not, for example, AU. The question is still open, but it is clear that any chemical specificity there may have been was not by itself sufficient to determine the code: the evolution of assignment enzymes remains the crucial step that has to be explained.

The basic idea (Szathmáry, 1993) is that the first involvement of amino acids in living processes was as cofactors of ribozymes. By recruiting amino acids cofactors, the catalytic range and efficiency of ribozymes could be greatly increased. The idea is shown in Figure 1. Each cofactor consisted of an amino acid bound to an oligonucleotide – probably a trinucleotide, in which case the code was a triplet code from the outset. The function of the oligonucleotide was to bind the cofactor to the ribozyme by base-pairing. Each type of cofactor could have acted in conjunction with many different riboyzmes.

In this scenario, the origin of the specific assignment of amino acids to oligonucleotides, which is the basis of the code, had at the outset nothing to do with protein synthesis. The assignments could have arisen one by one, each adding to the number of available cofactors, and hence to biochemical versatility. Subsequent evolutionary history is indicated in Figure 1. The next stage would have been the attachment to a single ribozyme of several amino acids. The linking of these to form peptides would then be the first step towards protein synthesis. Ultimately, the original ribozyme would evolve into mRNA; the oligonucleotide handle of the cofactor would evolve into tRNA; the assignment enzyme, R_2, attaching a particular amino acid to a particular oligonucleotide, would evolve into a tRNA-aminoacyl synthetase; and finally, the ribozyme R_3, linking amino acids into peptides, would evolve into the ribosome.

The model leaves many questions unanswered. For example, proteins are much larger than the short peptides that could be formed by using a ribozyme as a 'message'. But it has the advantage of suggesting intermediate stages between not having a code and having one, each of which could be selected for: for example, to have a single kind of cofactor would be better than to have none, to have two cofactors would be better than to have one, and so on. In this respect, it resembles other suggestions for the origins of complex organs that would seem useless until fully formed: for example,

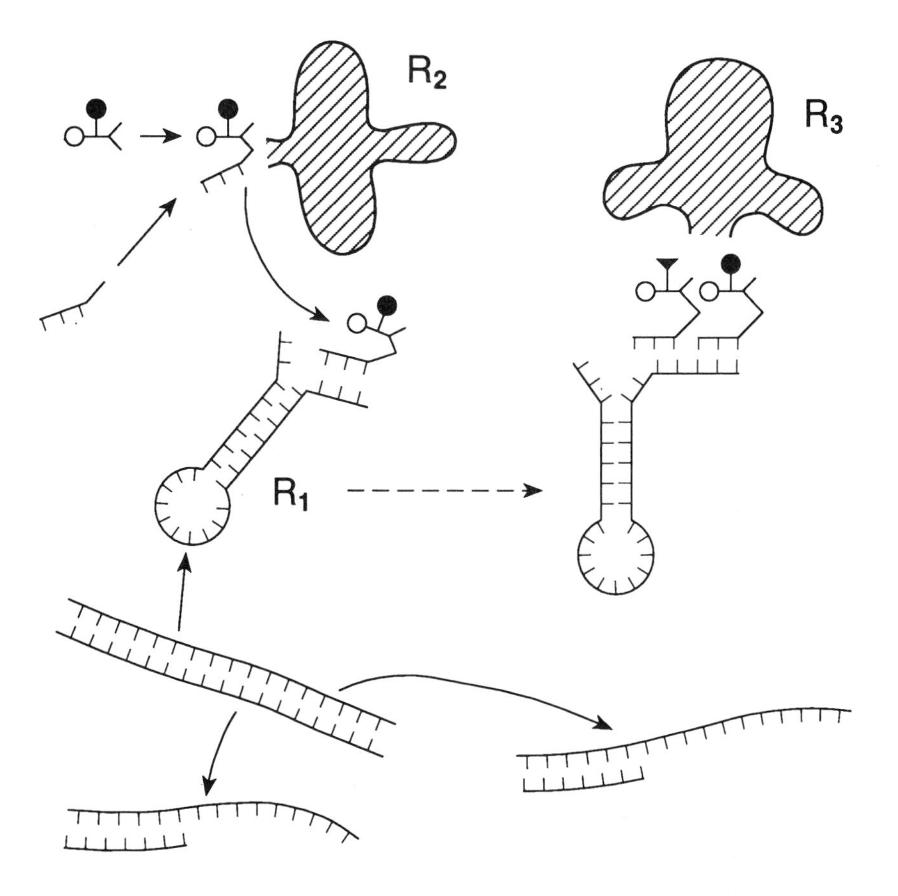

Figure 1. A hypothesis for the origin of the genetic code: for explanation, see text. → indicates changes within a cell; – – – –→ indicates evolutionary change.

feathers were useful in keeping their owners warm long before they were well enough formed to help in flight.

We now turn to our second problem, the origin of human language. This is a topic that has a bad reputation among linguists. After the publication of Darwin's *Origin of Species*, many uncritical ideas about the evolution of language were proposed, to such an extent that, in 1866, the French Academy of Linguistics announced that its journal did not accept papers on the origin of language. Their reaction was probably justified, but the time has come to reopen the question. In fact, in recent years there have been two exciting developments.

The first concerns the phylogeny of existing human languages. The phylogenetic approach to language is by no means new. To date, its main achievement has been the recognition that the Indo-European languages belong to a single family, with a common ancestor. Until recently, however, it was thought that the extent of word-borrowing between languages was so great that any attempt to discover a deeper phylogeny was hopeless. This view has now been challenged by a number of linguists, mainly from Russia and the United States. As so often in science, progress has depended on a refinement of methods. In this case the crucial step was the insistence that relationships should be deduced from shared vocabulary rather than grammar, which changes rather rapidly, and, more important, that the vocabulary should be confined to words without technical meanings: for example, words for parts of the body, relationships, sleeping and eating, hot and cold are suitable, but for ploughs, houses and arrows are not. The reason is obvious: technical words are more often borrowed.

It is interesting to compare phylogenetic reconstruction in biology and in language. Language reconstruction faces two main difficulties. The first, which arises particularly because we are concerned with spoken, not written, language, arises from sound shifts – for example, the systematic replacement in many words of the 'd' sound in German by 'th' in English. The nearest biological parallel is the change in the AT/GC ratio under mutational pressure. The second is word-borrowing. The parallel process in biology – horizontal gene transfer – has not often misled us. There is, however, one difficulty that is less severe in linguistics than in biology. In biology, particularly if we rely on morphological traits, we can be misled by convergence caused by similar selective forces in different lineages: the similarity between the eyes of a vertebrate and an octopus is an example. This difficulty is less severe in linguistics, because the forms of most words are unrelated to their meanings. Finally, as Cavalli-Sforza and his colleagues have shown, it is possible to check linguistic phylogenies by using genetic data. It may be too much to hope that we will be able to reconstruct the ur-language, or rather an ur-vocabulary, but real progress is being made in finding deeper relationships between languages.

This phylogenetic work is based on the assumption that all humans have a common competence for language: it is concerned with cultural, not biological, evolution. For a biologist, the more exciting question concerns the origin of linguistic competence itself. There has been a long debate between those who, following Skinner, see the learning of a language as just another example of human learning, achieved by suitable reinforcement – that is, by punishments and, more important, rewards – and those, following Chomsky, who argue that the ability to learn to speak is a peculiar one, and not merely a side effect of a general increase in intelligence. The latter argue that to talk requires an unconscious grasp of complex grammatical rules, which could not possibly be learnt in the way the behaviourists suggest.

It is now widely accepted that the Chomsky camp has won this debate. Two arguments have been decisive. The first points to the poverty of the input which is sufficient to enable a child to learn to talk. A child hears a finite set of sentences, but soon learns to utter an indefinitely large number. This implies that the child has learnt the rules whereby grammatically correct sentences can be generated, despite the fact that parents rarely correct the errors that a child makes. The second argument rests on the subtlety of the grammatical rules that must be learnt. Two generations of linguists and computer programmers have still not solved the problem of machine translation, yet many six-year-olds can speak two languages fluently, and translate from one to the other. Below we will discuss a third, genetic, argument for believing that humans possess a peculiar and innate competence for language.

It is easier to assert that linguistic competence is innate than to define precisely what that competence consists of. It seems that to produce and to comprehend speech depends on two abilities. The first is the ability to represent the meaning that is to be expressed in a hierarchical structure in the mind: the components of this structure are the elements that are represented, in the completed sentence, by noun phrases, verb phrases, and so on. The second is the ability to learn the rules whereby this semantic structure can be converted into a linear sequence of sounds – the 'surface structure'. The rules, of course, are different in different languages; for example, relationships that are conveyed by word order in English are conveyed by case endings in Latin. It is tempting to suggest that the first of these two abilities may have evolved because it served a cognitive rather than a communicating function. Thinking requires, not only the ability to form images in one's head, but to manipulate those images. To think 'two leopards climbed into that tree yesterday: one has come down, so there is still a leopard in the tree' would be a useful thing to be able to do, even if one could not express the thought in speech. It may be relevant that pre-linguistic children can perform comparable mental tasks. A linguist might object to this suggestion. Indeed, it is often argued that thinking can only be carried out in words. This seems doubtful. Playing chess, one might think 'If I play P×B, then he could play N–B3, forking my K and Q: so I must not take the bishop.' Although, unavoidably because we are trying to communicate, we have expressed this idea in words, the thought would be in visual images. But the thought is also grammatical, as the use of 'if', 'then' and 'so' makes clear. It is as if the nouns and verbs have been replaced by visual images, but the grammar remains. What grammar provides is the ability to carry out logical operations on images and concepts.

We suggest, therefore, that the ability to form concepts and to manipulate them evolved because thinking helped survival, independently of whether the thoughts could be communicated. The idea is not original: for example, Bickerton (1990), a linguist with evolutionary convictions, argues along these lines. It is hard to see, however, why the second ability, to convert the semantic structure into a linear sequence of sounds, should be needed except for communication. How could this competence evolve? Pinker and Bloom (1990) have argued that linguistic competence is a complex adaptive organ, in this sense resembling the eye of a vertebrate or the wing of a bird, and must therefore have evolved by natural selection. Although, as the authors themselves emphasize, the statement is obvious, it needed linguists to say it. The difficulty most linguists seem to have felt in imagining the origin of language lies in the difficulty of conceiving any useful intermediate between having language, and not having it. The difficulty is often expressed in the following form: if some grammatical rule – for example, the rule that converts a statement into a question – were absent, there would be important meanings that could not be expressed. Evolutionary biologists are familiar with this objection in other contexts. How often have we been told that the eye could not have evolved by natural selection because an eye that lacked some part, for example the iris, would not work? In the case of the eye, the objection can be answered, because there are surviving examples of light-sensitive organs with various intermediate degrees of complexity. The snag with language arises because such intermediates are lacking. It is hard enough to work out what the innate competence really is, without having to speculate about the intermediate stages it may have gone through during evolution.

Happily, a solution to this difficulty may come from an unexpected direction. Gopnik (1990; see also Gopnik & Crago, 1991) has described an English-speaking family, among whom a specific language disability is common. The disability occurs in 15 members of the family of 29 individuals, over three generations. It can occur in some but not all members of a sibship, so an environmental explanation – that children do not speak grammatically because one of their parents does not – is implausible. In fact, the condition is inherited as an autosomal dominant with high penetrance. It is specific both in the nature of the grammatical deficiency, described below, and in the fact that it is not associated with mental defect, deafness, motor impairment, or personality disorder: in particular, the affected children have an otherwise normal mental development.

Gopnik has used a number of tests to diagnose the condition, but its nature can best be explained by quoting some sentences written by affected children (we have slightly shortened some of these sentences, we hope without altering their significance):

'She remembered when she hurts herself the other day.'
'Carol is cry in the church.'
'On Saturday I went to nanny house with nanny and Carol.'

In each of these sentences, the child has failed to make an appropriate change in the form of a word: in the first two, a change is required to express the past tense (hurt, cried), and in the third to express possession (nanny's). Affected children have the same difficulty with plurals. A child will learn that a picture of a single book is a 'book', and of several books is 'books'. The child is then shown a picture of an imaginary animal, and is told that it is a 'wug': if then shown a picture of several wugs, the child does not know that the appropriate word is 'wugs'. Thus the child can learn particular examples of singular and plural, or of tense, just as we all have to learn the meanings of particular lexical items like 'horse' and 'cow', but does not generalize.

The failure to generalize is nicely illustrated by the following anecdote. Writing an account of what she did at the weekend, a child wrote:

'On Saturday I watch TV.'

Admittedly, this could be seen as a grammatically correct statement about what she is accustomed to doing on Saturdays. Reasonably, however, the teacher treated it as a statement about what she had done the previous weekend, and corrected it to 'watched'. Next weekend, the child wrote:

'On Saturday I wash myself and I watched TV and I went to bed.'

Three points emerge. She has learnt that the past tense of watch is watched. She has failed to generalize to washed. She already knows that the past of go is went: after all, this is something we must all learn as a unique fact, not by generalization.

This fascinating case has some important implications. First, although the affected people are impaired, they are not without grammar: they are a great deal better off than they would be if they could not talk at all. In other words, there can be intermediates between no competence and perfect competence. Second, the impairment is specific to language: there is no mental defect. This confirms Chomsky's view that linguistic competence is not a mere spin-off of general intelligence. Third, it suggests a road to understanding the evolution of language.

If, as seems very likely, the impairment is caused by a mutation in a single autosomal gene, the possibility exists that the gene could be located and characterized. It is not clear what such a characterization would tell us. If there is one such gene, there must surely be others, although, if the mutations are recessive or of imperfect penetrance, they will be harder to find. It is already known that specific language impairment is not confined to this one family. In an otherwise excellent review of its epidemiology, Tomblin (in preparation) makes what may prove to be the misleading assumption that specific language impairment is a single entity. It is worth remembering how important was the recognition by Penrose (1949) that the term 'mental defect' was being used to cover a number of genetically distinct conditions. We can expect that, in the next ten years, a number of different gene loci, each with a different effect on linguistic competence, may be discovered.

What, if anything, will that tell us about the nature of linguistic competence? Perhaps we should not be too optimistic. For over fifty years, geneticists have believed that the study of genes with specific effects on development was the best road to understanding how development works. Until very recently, the belief had rather little to justify it. Now, studies of *Drosophila, Caenorhabditis, Arabidopsis*, and the mouse seem at least to be yielding the promised fruit. A genetic dissection of grammar is likely to be much harder, partly because we do not know clearly what we are trying to explain, and partly because experiments can be performed on fruit-flies that cannot be performed on children. But despite these grounds for caution, the prospect of collaboration between linguists and geneticists, after a long period of mutual distrust, is very exciting.

REFERENCES

Bickerton, D. (1990). *Language and Species*. Chicago: University of Chicago Press.

Gopnik, M. (1990). Feature-blind grammar and dysphasia. *Nature* **344**, 715.

Gopnik, M. & Crago, M. B. (1991). Familial aggregation of a developmental language disorder. *Cognition* **39**: 1–50.

Osawa, S., Jukes, T. H., Watanabe, K. & Muto, A. (1992). Recent evidence for evolution of the genetic code. *Microbiological Reviews* **56**, 229–264.

Penrose, L. S. (1949). *The Biology of Mental Defect*. London: Sidgwick & Jackson.

Pinker, S. & Bloom, P. (1990). Natural language and natural selection. *The Brain and Behavioural Sciences* **13**, 707–784.

Szathmáry, E. (1993). Coding coenzyme handles: a hypothesis for the origin of the genetic code. *Proceedings of the National Academy of Sciences USA* **90**, 9916–9920.

Tomblin, J. B. (in preparation). Epidemiology of specific language impairment. In *Biological Aspects of Language*, ed. M. Gopnik. Oxford: Oxford University Press.

Zaug, A. J. & Cech, T. R. (1986). The intervening sequence of tRNA of *Tetrahymena* is an enzyme. *Science* **231**, 470–475.

7

RNA without protein or protein without RNA?

CHRISTIAN DE DUVE

International Institute of Cellular and Molecular Pathology, Brussels, and The Rockefeller University, New York

The answer to the question posed in the title of this communication depends on what is meant by 'protein'. If we restrict this term to polypeptides assembled on ribosomes from a set of twenty tRNA-linked L-amino acids according to an mRNA sequence, then we may safely assume that RNA preceded protein in the development of life, since all the main components of the protein-synthesizing machinery are RNA molecules. Such is the view embodied in the now widely accepted model of an 'RNA world' (Crick, 1968; Gilbert, 1986). On the other hand, if we enlarge the definition of protein to include any kind of polypeptide, then there is a good possibility that protein may have preceded RNA, since amino acids were probably among the most abundant biogenic building blocks available on the prebiotic Earth (Miller, 1992), and their spontaneous polymerization, although not readily accounted for in an aqueous medium, is at least easier to visualize than the spontaneous assembly of RNA molecules. Let us consider first proteins *stricto sensu*. How did such molecules come into being?

According to the most reasonable hypothesis (Orgel, 1989; de Duve, 1991; 1995), primary interactions between amino acids and RNA molecules led to the progressive assembly of a primitive, as yet uninformed peptide-synthesizing machinery. Subsequent evolution of this system saw the gradual development of translation and of the genetic code. This long evolutionary process must have been driven first by the enhanced replicatability/stability of the RNA molecules involved. Later, as translation fidelity improved, advantageous properties of the synthesized peptides became increasingly important. Eventually, the properties of the peptides dominated the evolutionary process. Among these properties, catalytic activities no doubt played a major role. It is likely that polypeptide enzymes first appeared in the course of this process and were selected on the strength of their ability to catalyse some chemical reaction. Such a selection mechanism holds an interesting implication.

Let us assume a mutation leading to the formation of an enzyme catalysing the conversion of A to B. Obviously, such an enzyme could have been of no use if A had not been present. It would have been of little use also if there had not been an outlet for B. Extending this reasoning to every new enzyme that appeared as a result of some mutation and was retained by natural selection, we arrive at the conclusion that many of the substrates and products of these enzymes must have pre-existed in the RNA world. I see in this a strong argument in support of the contention (de Duve, 1993; 1995) that protometabolism – the set of chemical reactions that generated and supported the RNA world – and metabolism – the set of enzyme-catalysed reactions that support present-day life – must have been largely *congruent*, that is, must have followed largely similar pathways.

This conclusion is relevant to the central question of how RNA first arose. In spite of considerable effort, no plausible answer to this question has yet been found (Joyce, 1991). The possibility of some fluke event or random fluctuation, somehow perpetuated by replication, cannot be contemplated. We are dealing with a robust set of reactions, the core of protometabolism, the underpinning of the RNA world for all the time it took enzyme-catalysed metabolism to develop. The congruence argument suggests that one should look more closely at the biological pathways of RNA synthesis to understand the prebiotic formation of this central substance. This view conflicts with the commonly accepted notion that prebiotic mechanisms must have been very different from metabolic mechanisms. I believe, however, that the congruence argument is not easily refutable.

The concept of an abiotic chemistry unrelated to biochemistry rests on the consideration that metabolism depends on the catalytic activities of protein enzymes that could not have been present on the prebiotic Earth. Hence the necessity of identifying reactions able to proceed without catalysts or with the sole help of mineral catalysts. This point, however, applies only to whatever steps were needed for abiotic chemistry to produce its own catalysts. There is no reason for assuming that ribozymes were the first biological catalysts. The possibility that peptide catalysts arose earlier is perfectly conceivable and is, in fact, more probable in terms of chemical feasibility. Furthermore, peptide catalysts seem most likely to have possessed activities similar to those of present-day enzymes, as required by the congruence principle.

Unlike ribozymes, peptide catalysts could not have been replicated – at least if Crick's 'Central Dogma' was already valid four billion years ago – and, therefore, could not have been subject to selection by way of mutation. This, however, is true of any other pre-ribozyme catalyst – unless one accepts an involvement of replicatable mineral catalysts (Cairns-Smith, 1982) – and can hardly be raised as an objection against the participation of peptide catalysts in protometabolism. All that would have been needed was a stable and reproducible supply of peptides containing all the required catalysts. The condition of stability and reproducibility could have been satisfied by a stable set of environmental conditions. As to the condition of catalytic sufficiency, there are good reasons for believing that relatively simple peptides would already be endowed with catalytic activities. This assumption is supported by what is known of the modular construction of proteins and by what is suspected of the size of the first genes and their products. According to Eigen, in order to be replicatable without irretrievable loss of information, the first RNA genes could not have been more than 70–100 nucleotides long (Eigen *et al.*, 1981) which means that the first translation products, among which the first enzymes presumably were present, were no more than 20–30 amino acid residues long.

As to the prebiotic formation of peptides, it raises a problem common to all prebiotic condensation reactions. Two solutions to this problem exist in principle. Either condensation took place in the absence of water, as in Fox's thermal synthesis of 'proteinoids' (Fox & Harada, 1958). Or, alternatively, some condensing or activating agent was available. The possible involvement of pyrophosphate or of some polyphosphate has often been evoked. My preference goes to thioesters (de Duve, 1991). The thioester bond plays a central and, most likely, very ancient role in energy metabolism. In addition, a number of bacterial peptides are actually synthesized from the thioesters of amino acids in the present-day world (Kleinkauf & von Döhren, 1987). This reaction can be reproduced in the absence of catalyst under very simple conditions (Wieland, 1988).

Whatever the mechanisms involved, I believe the case for congruence and for an intervention of peptide catalysts in the pre-RNA world rests on a sound theoretical basis. It could be tested in the laboratory with randomly synthesized peptide mixtures. Primitive enzyme-like catalysts should be detectable in such mixtures.

REFERENCES

Cairns-Smith, A. G. (1982). *Genetic Takeover and the Mineral Origins of Life.* Cambridge: Cambridge University Press.

Crick, F. H. C. (1968). The origin of the genetic code. *Journal of Molecular Biology* **38**, 367–379.

de Duve, C. (1991). *Blueprint for a Cell.* Burlington, NC: Neil Patterson Publishers, Carolina Biological Supply Company.

de Duve, C. (1993). The RNA world: before and after? *Gene* **135**, 29–31.

de Duve, C. (1995). *Vital Dust: Life as a Cosmic Imperative.* New York: Basic Books.

Eigen, M., Gardiner, W., Schuster, P. & Winkler-Oswatitsch, R. (1981). The origin of genetic information. *Scientific American* **244**, 88–118.

Fox, S. W. & Harada, K. (1958). Thermal copolymerisation of amino acids in a product resembling protein. *Science* **128**, 1214.

Gilbert, W. (1986). The RNA world. *Nature* **319**, 618.

Joyce, G. F. (1991). The rise and fall of the RNA world. *New Biologist* **3**, 399–407.

Kleinkauf, H. & von Döhren, H. (1987). Biosynthesis of peptide antibiotics. *Annual Reviews of Microbiology* **41**, 259–289.

Miller, S. L. (1992). The prebiotic synthesis of organic compounds as a step toward the origin of life. In *Major Events in the History of Life*, ed. J. W. Schopf, pp. 1–28. Boston, MA: Jones and Bartlett.

Orgel, L. E. (1989). The origin of polynucleotide-directed protein synthesis. *Journal of Molecular Evolution* **29**, 465–474.

Wieland, T. (1988). Sulfur in biomimetic peptide synthesis. In *The Roots of Modern Biochemistry*, eds. H. Kleinkauf, H. von Döhren & L. Jaenicke, pp. 213–221. Berlin: Walter de Gruyter.

8

'What is life?': was Schrödinger right?

STUART A. KAUFFMAN
Sante Fe Institute, New Mexico

In Dublin half a century ago, a major figure in this century's science visited, lectured, and foretold the future of a science which was not his own. The resulting book, *What is Life?*, is credited with having inspired some of the most brilliant minds ever to enter biology to the work which gave birth to molecular biology (Schrödinger, 1944). Schrödinger's 'little book' is, itself, as brilliant as warranted by its reputation. But, half a century later, and at the occasion of its honouring, perhaps we may dare to ask a new question: is the central thesis of the book right? I mean no dishonour to so superb a mind as Schrödinger's, nor to those properly inspired by him, to suggest that he may have been wrong, or at least incomplete. Rather, of course, like all scientists inspired by his ideas, I too seek to continue the quest.

I am hesitant even to raise the questions I shall raise, for I am also fully aware of how deeply embedded Schrödinger's own answers are in our view of life since Darwin and Weismann, and since the development of the theory of the germ plasma, with the gene as the necessary stable storage form of heritable variation: 'Order from order', answered Schrödinger. The large aperiodic solids and the microcode of which Schrödinger spoke have become the DNA and the genetic code of today. Almost all biologists are convinced that such self-replicating molecular structures and such a microcode are essential to life.

I confess I am not entirely convinced. At its heart, the debate centres on the extent to which the sources of order in biology lie predominantly in the stable bond structures of molecules, Schrödinger's main claim, or in the collective dynamics of a system of such molecules. Schrödinger emphasized, correctly, the critical role played by quantum mechanics, molecular stability, and the possibility of a microcode directing ontogeny. Conversely, I suspect that the ultimate sources of self-reproduction and the stability requisite for heritable variation, development and evolution, while requiring the stability of organic molecules, may also require emergent ordered properties in the collective behaviour of complex, non-equilibrium chemical reaction systems. Such complex reaction systems, I shall suggest, can spontaneously cross a threshold, or phase transition, beyond which they become capable of collective self-reproduction, evolution, and exquisitely ordered dynamical behaviour. The ultimate sources of the order requisite for life's emergence and evolution may rest on new principles of collective emergent behaviour in far from equilibrium reaction systems.

In brief, foreshadowing what follows: while Schrödinger's insights were correct about current life, I suspect that in a deeper sense, he was incomplete. The formation of large aperiodic solids carrying a microcode, order from order, may be neither necessary nor sufficient for the emergence and evolution of life. In contrast, certain kinds of stable collective dynamics may be both necessary and sufficient for life. I would emphasize that I raise these issues for discussion, not as established conclusions.

SCHRÖDINGER'S ARGUMENT

Schrödinger begins his discussion by emphasizing the view of macroscopic order held by most physicists of his day and earlier. Such order, he tells us, consists in averages over enormous ensembles of atoms or molecules. Statistical mechanics is the proper intellectual framework for this analysis. Pressure in a gas confined in a volume is just the average behaviour of very large numbers of molecules colliding with and recoiling from the walls. The orderly behaviour is an average, and not due to the behaviour of individual molecules.

But what accounts for the order in organisms, and, in particular, for rare mutations and heritable variation? Schrödinger then uses current data to estimate the number of atoms which might be involved in a gene, and correctly estimates that the number cannot be more than a few thousand atoms. Order due to statistical averaging cannot help here, he argues, for the numbers of atoms are too small for reliable behaviour. In statistical systems the expected sizes of the fluctuations vary inversely with the square root of the number of events. With ten tosses of a fair coin, 80% 'heads' is not surprising, with ten thousand tosses, 80% 'heads' would be stunning.

With a million events, Schrödinger points out, statistical fluctuations would be on the order of 0.001, rather unreliable for the order found in organisms.

Quantum mechanics, argues Schrödinger, comes to the rescue of life. Quantum mechanics ensures that solids have rigidly ordered molecular structures. A crystal is the simplest case. But crystals are structurally dull. The atoms are arranged in a regular lattice in three dimensions. If you know the positions of the atoms in a minimal 'unit crystal', you know where all the other atoms are in the entire crystal. This overstates the case, of course, for there can be complex defects in crystals, but the point is clear. Crystals have very regular structures, so the different parts of the crystal, in some sense, all 'say' the same thing. In a moment Schrödinger will translate the idea of 'saying' into the idea of 'encoding'. With that leap, a regular crystal cannot encode much information. All the information is contained in the unit cell.

If solids have the order required, but periodic solids such as crystals are too regular, then Schrödinger puts his bet on aperiodic solids. The stuff of the gene, he bets, is some form of aperiodic crystal. The form of the aperiodicity will contain some kind of microscopic code which, somehow, controls the development of the organism. The quantum character of the aperiodic solid will mean that small discrete changes, mutations, will occur. Natural selection, operating on these small discrete changes, will select out favourable mutations as Darwin hoped.

Schrödinger was right. His book deserves its fine reputation. Five decades later we know the structure of DNA. There is, indeed, a code leading from DNA to RNA and to the primary structure of proteins. This would have been a wonderful success for any scientist, let alone a physicist peering over the wall at biology.

But is Schrödinger's insight either necessary or sufficient? Is the order assembled in the DNA aperiodic crystal either necessary or sufficient for the evolution of life, or for the dynamical order found in current life? Neither, I suspect. The ultimate sources of order may require the discrete order of stable chemical bonds derived from quantum mechanics, but lie elsewhere. The ultimate sources of order and self-reproduction may lie in the emergence of collectively ordered dynamics in complex chemical reaction systems.

The main part of this chapter has two sections. The first examines briefly the possibility that the emergence of life itself is not based on the template replicating properties of DNA or RNA, but on a phase transition to collectively autocatalytic sets of molecules in open thermodynamic systems. The second section examines the emergence of collective dynamical order in complex parallel processing networks of elements. Those elements might be genes whose activities are mutually regulated, or might be the polymer catalysts in an autocalytic set. Such networks are open thermodynamically, and the core source of the dynamical order they exhibit lies in the way dynamical trajectories converge to small attractors in the phase space of the system.

Since I want to suggest that convergence to small attractors in open thermodynamic systems is a major source of order in living organisms, I want to end this introductory section by laying out the background of Schrödinger's discussion about statistical laws.

The central point is simple: in closed thermodynamic systems, there is no convergence in the appropriate phase space. The character of the resulting statistical laws reflects this lack of convergence. But in some open thermodynamic systems there can be massive convergence of the dynamical flow of the system in its state space. This convergence can engender order rapidly enough to offset the thermal fluctuations which always occur.

The critical distinction between a closed system at equilibrium and an open system displaced from equilibrium is this: in a closed system, no information is thrown away. The behaviour of the system is, ultimately, reversible. Because of this, phase volumes are conserved. In open systems, information is discarded into the environment and the behaviour of the subsystem of interest is not reversible. Because of this, the phase volume of the subsystem can decrease. I am no physicist, but will try to lay out the issues simply, and I hope, correctly.

Consider a gas confined to a box, closed to exchange of matter and energy. Every possible microscopic arrangement of the gas molecules is as likely as any other. The motions of the molecules are governed by Newton's laws. Hence the motions are microscopically reversible, and the total energy of the system is conserved. When molecules collide, energy is exchanged but not lost. The 'ergodic hypothesis', something of a leap of faith that works, asserts that as these molecular collisions occur, the total system visits all possible microstates over time equally often. Thus, the probability that the system is in any macrostate is exactly equal to the fractional number of microstates corresponding to that macrostate.

Liouville's theorem states that volumes in phase space are conserved under the flow of an equilibrium system. For a system with N gas molecules, the current position and momentum of each molecule in three spatial dimensions can be represented by six numbers. Hence in a $6N$-dimensional phase space, the current state of the entire volume of gas can be represented as a single point. Consider a set of nearly identical initial states of the gas in a box. The corresponding points occupy some volume in phase space. Liouville's theorem asserts that, as molecular collisions occur in each copy of the box of gas, the corresponding volume in phase space moves, deforms, and smears out over the phase space. But the total volume in phase space remains constant. There is no convergence in the flow in phase space. Since phase volume is constant, then given the ergodic hypothesis, the probabilities of macrostates are just proportional to the relative numbers of microstates within each macrostate, normalized by the total number of microstates.

Suppose, instead, that the flow of the system in phase space allowed the initial phase volume to progressively contract to a single point, or to a small volume. Then the spontaneous behaviour of the system would flow to some unique configuration, or some small number of configurations. Order would emerge! Of course, such convergence cannot happen in a closed, equilibrium thermodynamic system. If it did, entropy would decrease rather than increase in the total system.

Such order can emerge. Clearly the emergence of this kind of order requires as a necessary condition that the system be thermodynamically open to the exchange of matter and energy. This exchange allows information to be lost from the subsystem of interest into its environment. A physicist would say that 'degrees of freedom' – the diverse ways the molecules can move and interact – are lost into the heat bath of the environment.

Thus, the kind of dynamical order we seek can only arise in non-equilibrium thermodynamic systems. Such systems have been termed dissipative structures by Prigogine. Whirlpools, Zhabotinsky reactions, Bénard cells, and other examples are now familiar. However, it is essential to stress that discplacement from thermodynamic equilibrium by itself is only a necessary condition, not a sufficient condition, for the emergence of highly ordered dynamics. The fabled butterfly in Rio de Janeiro whose wings beget chaos in the weather can recur in many versions in complex, non-equilibrium chemical systems, begetting chaos which would forbid the emergence and evolution of life. In the third section of this chapter I return to the emergence of collectively ordered dynamical behaviour in non-equilibrium open systems.

THE ORIGIN OF LIFE AS A PHASE TRANSITION

The large aperiodic solid about which Schrödinger mused is now known. Since Watson and Crick remarked, with uncertain modesty, that the template complementary structure of the DNA double helix foretold its mode of replication, almost all biologists have fastened upon some version of template complementarity as necessary to the emergence of self-reproducing molecular systems. The current favourite contenders are either RNA molecules or some similar polymer. The hope is that such polymers might act as templates for their own replication, in the absence of any outside catalyst.

So far, efforts to achieve replication of RNA sequences in the absence of enzymes have met with limited success. Leslie Orgel, a speaker at the Trinity College, Dublin conference and an outstanding organic chemist, has worked hard to achieve such molecular replication (Orgel, 1987). He, better than I, could summarize the difficulties. But, briefly, the problems are many. Abiotic synthesis of nucleotides is difficult to achieve. Such nucleotides like to bond by 2′–5′ bonds rather than the requisite 3′–5′ bonds. One wants to find an arbitrary sequence of the four normal RNA bases and use these as a template to line up the four complementary bases by hydrogen bond pairing, such that the lined up nucleotides form the proper 3′–5′ bonds and melt from the initial template, and the system cycles again to make an exponential number of copies. It has not worked as yet. The difficulties arise for good chemical reasons. For example, a single strand which is richer in C than G will form the second strand as hoped. But the second, richer in G than C, tends to form G–G bonds which cause the molecule to fold in ways which preclude function as a new template.

With the discovery of ribozymes and the hypothesis of an RNA world, a new, attractive hope is that an RNA molecule might function as a polymerase, able to copy itself and any other RNA molecule. Jack Szostak, at Harvard Medical School, is attempting to evolve such a polymerase *de novo.* If he succeeds, it will be a *tour de force.* But I am not convinced that such a molecule holds the answer to life's emergence. It seems a rare structure to have been formed by chance as the first 'living molecule'. And, were it formed, I am not yet persuaded it could evolve. Such a molecule, like any enzyme, would make errors when replicating itself, and hence form mutant copies. These would compete with the 'wild type' ribozyme polymerase to replicate that polymerase, and would themselves tend to be more error prone. Hence the mutant RNA polymerases would tend to generate still more badly mutant RNA polymerase sequences. A runaway 'error catastrophe', of a type first suggested by Orgel concerning coding and its translation, might occur. I do not know that such an error catastrophe would arise, but believe the problem is worth analysis.

When one contemplates the symmetrical beauty of the DNA double helix, or a similar RNA helix, one is forced to admit the simple beauty of the corresponding hypothesis. Surely such structures were the first living molecules. But is this really true? Or might the roots of life lie deeper. I turn next to explore this possibility.

The simplest free living organisms are the mycoplasma. These derived bacterial forms have on the order of 600 genes encoding proteins via the standard machinery. Mycoplasma possess membranes, but no bacterial cell wall. They live in very rich environments, the lungs of sheep or man, for example, where their requirements for a rather large variety of exogenous small molecules are met.

Why should the simplest free living entities happen to harbour something like 600 kinds of polymers and a metabolism with perhaps 1000 small molecules? And how, after all, does the mycoplasm reproduce? Let us turn to the second question first, for the answer is simple and vital. The mycoplasma cell reproduces itself by a kind of collective autocatalysis. No molecular species within the mycoplasma actually replicates itself. We know this, but tend to ignore it. The DNA of the mycoplasma is replicated thanks to the coordinated activities of a host of cellular enzymes. In turn, the latter are synthesized via standard messenger RNA sequences. As we all know, the code is translated from RNA to proteins only with the help of encoded proteins – namely the amino acid synthetases which charge each transfer RNA properly for later assembly by the ribosome into a protein. The membrane of the cell has its molecules formed by catalysis from metabolic intermediates. We are all familiar with the story. No molecule in the mycoplasm replicates itself. The system as a whole is collectively autocatalytic. Every molecular species has its formation catalysed by some molecular species in the system, or else is supplied exogenously as 'food'.

If the mycoplasma is collectively autocatalytic, so too are all free living cells. In no cell does a molecule actually replicate itself. Then let us ask why the minimal complexity found in free living cells is on the order of 600 protein polymers and about 1000 small molecules. We have no answer. Let us ask why there should be a minimal complexity under the standard hypothesis that single stranded RNA sequences might serve as templates and be able to replicate without other enzymes. But on this hypothesis there could be no deep answer to why a minimal complexity is observed in all free living cells. Just the simplicity of the 'nude' replicating gene is what commends that familiar hypothesis to us. All we might respond is that, 3.45 billion years later, the simplest free living cells happen to have the complexity of mycoplasma. We have no deep account, merely another evolutionary 'Just So' story.

I now present, in brief form, a body of work carried out over the past eight years alone and with my colleagues (Kauffman, 1971, 1986, 1993; Farmer *et al.*, 1986; Bagley, 1991; Bagley *et al.*, 1992). Similar ideas were advanced independently by Rossler (1971), Eigen (1971), and Cohen (1988). The central concept is that, in sufficiently complex chemical reaction systems, a critical diversity of molecular species is crossed. Beyond that diversity, the chances that a subsystem exists which is collectively autocatalytic goes to 1.0.

The central ideas are simple. Consider a space of polymers including monomers, dimers, trimers, and so forth. In concrete terms, the polymers might be RNA sequences, peptides or other kinds of polymeric forms. Later, the restriction to polymers will be removed and we will consider systems of organic molecules.

Let the maximum length of polymer we will think about be M. And let M increase. Count the number of polymers in the system, from monomers to polymers of length M. It is simple to see that the number of polymers in the system is an exponential function of M. Thus, for 20 kinds of amino acids, the total diversity of polymers up to length M is slightly greater than 20^M. For RNA sequences, the total diversity is slightly more than 4^M.

Now let us consider all cleavage and ligation reactions among the set of polymers up to length M. Clearly, an oriented polymer such as a peptide or RNA sequence of length M can be made from smaller sequences in $M - 1$ ways, since any internal bond in the polymer of length M is a site at which smaller fragments can be ligated. Thus, in the system of polymers up to length M, there are exponentially many polymers, but there are even more cleavage and ligation reactions by which these polymers can interconvert. Indeed, as M increases, the ratio of cleavage and ligation reactions per polymer increases linearly with M.

Define a reaction graph among a set of polymers. In general, we might think of one substrate one product reactions, one substrate two product reactions, two product one substrate reactions, and two substrate two product reactions. Transpeptidation and transesterification reactions are among the two substrate two product reactions that peptides or RNA sequences can undergo. A reaction graph consists of the set of substrates and products, which may be pictured as points, or nodes scattered in a three-dimensional space. In addition, each reaction can be denoted by a small circular 'reaction box'. Arrows from the substrate(s) leads to the box. Arrows lead from the box to the product(s). Since all reactions are actually at least weakly reversible, the arrow directions are merely useful to indicate which sets of molecules constitute the substrates and which the products in one of the two directions the reaction may take. The reaction graph consists in this entire collection of nodes, boxes, and arrows. It shows all possible reactions among the molecules in the system.

The implication noted above of the combinatorics of chemistry is that, as the diversity of polymers in the system increases, the ratio of reactions to molecules increases. This means that the ratio of arrows and boxes to nodes increases. The reaction graph becomes ever denser, ever more richly interconnected with reaction possibilities, as the diversity of molecules in the system increases.

In such a reaction system it is always true that some reactions happen spontaneously at some velocity. I ask the reader to ignore, for the moment, such spontaneous reactions in order to focus on the following question: *under what conditions will a collectively autocatalytic set of molecules emerge?* I aim to show that, under a wide variety of hypotheses about the system, autocatalytic sets will emerge at a critical diversity.

I begin by drawing attention to well known phase transitions in random graphs. Throw ten thousand buttons on the floor and begin to connect pairs of buttons at random with red threads. Such a collection of buttons and threads is a random graph. More formally, a random graph is a set of nodes connected at random with a set of edges. Every so often pause to hoist a button and see how many buttons one lifts up with it. Such a connected set of buttons is called a component in a random graph. Erdos and Renyi showed, some decades ago (1960), that such systems undergo a phase transition as the ratio of edges to nodes passes 0.5. When the ratio is lower, when the number of edges is, say 10% of the number of nodes, any node will be directly or indirectly connected to only a few other nodes. But when the ratio of edges to nodes is 0.5, suddenly most of the nodes become connected into a single giant component. Indeed, if the number of nodes were infinite, then as the ratio of edges to nodes passed 0.5 the size of the largest component would jump discontinuously from very small to infinite. The system exhibits a first order phase transition. The point to bear away is simple: when enough nodes are connected, even at random, a giant interconnected component literally crystallizes.

We need only apply this idea to our reaction graph. Looking ahead, we will focus attention on catalysed reactions. We will need a theory about which polymers catalyse which reactions. Given a variety of such theories, we will find a simple consequence: as the diversity of molecules in the system increases, the ratio of reactions to molecules increases. Thus, for almost any model of which polymers catalyse which reactions, at some diversity almost every polymer will catalyse at least one reaction. At that critical diversity a giant component of connected catalysed reactions will crystallize in the system. If the polymers which act as catalysts are themselves the products of the catalysed reactions, the system will become collectively autocatalytic.

But this step is easy. Consider a simple, indeed oversimple, model of which polymer catalyses which reaction. I will relax the idealization I am about to make further below. Let us assume that any polymer has a fixed probability, say one in a billion, to be able to act as a catalyst for any randomly chosen reaction. Now consider our reaction graph at a point when the diversity of molecules in the system is such that there are a billion reactions for every molecule. And let the molecules in question be polymers which are themselves candidates to catalyse the reactions among the polymers. But then about one reaction per polymer will be catalysed. A giant component will crystallize in the system. And, with a little thought, it is clear that the system almost certainly contains collectively autocatalytic subsystems. Self-reproduction has emerged at a critical diversity, owing to a phase transition in a chemical reaction graph.

Figure 2 shows such a collectively autocatalytic set. The main point to stress is the near inevitable emergent property of such systems, and a kind of unrepentant holism. At a lesser diversity, the resulting reaction graph has only a few reactions which are catalysed by molecules in the system. No autocatalytic set is present. As the diversity increases, a larger number of reactions are catalysed by molecules in the system. At some point as the diversity increases, a connected web of catalysed reactions springs into existence. The web embraces the catalysts themselves. Catalytic closure is suddenly attained. A 'living' system, self-reproducing at least in its silicon realization, swarms into existence.

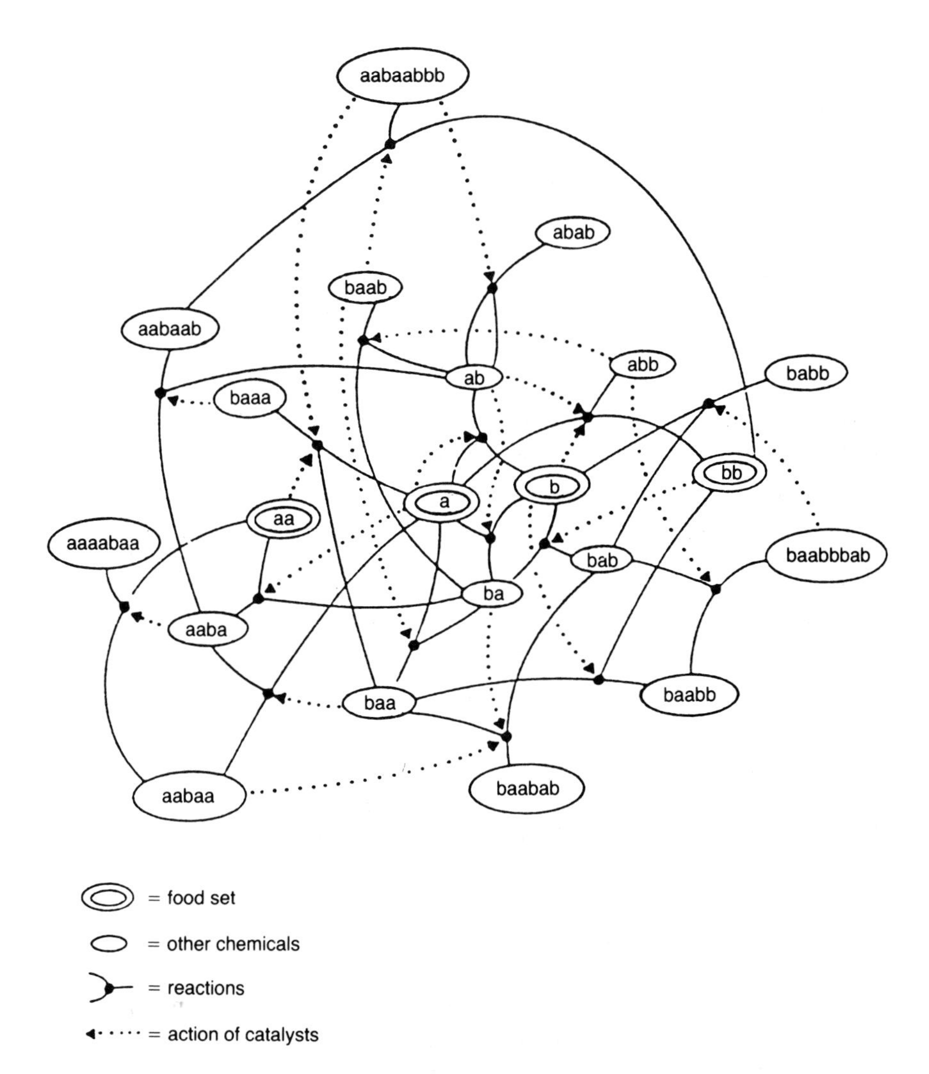

Figure 2. A typical example of a small autocatalytic set. The reactions are represented by points connecting cleavage products with the corresponding larger ligated polymer. Dotted lines indicate catalysis and point from the catalysts to the reaction being catalysed. Monomers and dimers of A and B constitute the maintained food set (double ellipses).

Further, this crystallization requires a critical diversity. A simpler system simply does not achieve catalytic closure. We begin to have a candidate for a deep theory for the minimal diversity of free living cells. No 'Just So' story here: simpler systems fail to achieve or sustain autocatalytic closure.

The total molecular diversity required to cross the phase transition depends upon two major factors: (1) the ratio of reactions to molecules, and (2) the distribution of probabilities that the molecules in the system catalyse the reactions among the same molecules. The ratio of reactions to molecules depends upon the complexity of the kinds of reactions allowed. For example, if one considers only cleavage and ligation reactions among peptides or RNA sequences, then the ratio of reactions to polymers increases linearly with maximum length, M, of polymer in the system. This is easy to see in outline, since a polymer of length M can be made in $M - 1$ ways. As M increases, the ratio of reactions to polymers increases proportionally. Conversely, one might consider transpeptidation or transesterification reactions among peptides or RNA sequences. In that case the ratio of reactions to polymers increases very much faster than linearly. Consequently, the diversity of molecules requisite for the emergence of autocatalytic sets is much smaller. In concrete terms, if the probability an arbitrary polymer catalysed an arbitrary reaction were one in a billion, then about 18 000 kinds of molecules would suffice for the emergence of collective autocatalytic sets.

The results we are discussing are robust with respect to the over simple idealization that any polymer has a fixed probability of being able to function as a catalyst for any reaction. An alternative model (Kauffman, 1993) considers RNA sequences as the potential, simple ribozymes, and supposes that in order to function as a specific ligase, a candidate ribozyme must template match the three terminal 5′ nucleotides on one substrate and the three terminal 3′ nucleotides on a second substrate. Recently, von Kiederowski (1986) has generated just such specific ligases which actually form small autocatalytic sets! A hexamer ligates two trimers which then constitute the hexamer. More recently, von Kiederowski has created collectively reproducing cross-catalytic systems (personal communication, 1994). In line with von Kiederowski's results, in our model RNA system to capture the fact that other features beyond template matching may be required for the candidate RNA to actually function as a catalyst in the reaction, Bagley and I assumed that any such matching candidate still had only a one in a million chance to be able to function as a specific ligase. Collectively autocatalytic sets still emerge at a critical diversity of model RNA sequences in the system. Presumably, the results are robust and will remain valid for a wide variety of models about the distribution of catalytic capacities among sets of polymers or other organic molecules. I return in a moment to discuss experimental avenues to attempt to create such collectively autocatalytic systems.

If this view is right, then the emergence of life does not depend upon the beautiful templating properties of DNA, RNA, or other similar polymers. Instead, the roots of life lie in catalysis itself and in the combinatorics of chemistry. If this view is right, then the routes to life may be broad boulevards of probability, not back alleys of rare chance.

But can such collectively autocatalytic systems evolve? Can they evolve without a genome in the familiar sense? And if so, what are the implications for our tradition since Darwin, Weismann, and, indeed, since Schrödinger? For, if self-reproducing systems can evolve without a stable large molecular repository of genetic information, then Schrödinger's suggestion about large aperiodic solids is not necessary to the emergence and evolution of life.

At least in computer experiments such collectively autocatalytic systems can evolve without a genome. First, I should stress that my colleagues, Farmer and Packard, and I (1986), have shown, using fairly realistic thermodynamic conditions in model stirred flow reactors, that model autocatalytic systems can, in fact, emerge. Further, Bagley has shown as part of his thesis that such systems can attain and sustain high concentrations of large model polymers in the face of a bias towards cleavage in an aqueous medium. Moreover, such systems can 'survive' if the 'food' environment is modified in some ways, but are 'killed' – i.e. collapse – if other foodstuffs are removed from the flow reactor system. Perhaps the most interesting results, however, show that such systems can evolve to some extent without a genome. Bagley *et al.* (1992), made use of the reasonable idea that spontaneous reactions which persist in the autocatalytic set will tend to give rise to molecules which are not members of the set itself. Such novel molecules form a kind of penumbra of molecular species around the autocatalytic set, and are present at higher concentrations than they would otherwise be, owing to the presence of the autocatalytic set. The autocatalytic set can evolve by grafting some of these new molecular species into itself. It suffices if one or more of these shadow set molecules fluctuates to a modest concentration, and that these molecules then aid catalysts of their own formation from the autocatalytic set. If so, the set expands to include these new molecular species. Presumably, if some molecules can inhibit reactions catalysed by other molecules, the addition of new kinds of molecules will sometimes lead to the elimination of older kinds of molecules.

In short, at least *in silico*, autocatalytic sets can evolve without a genome.

No stable large molecular structure carries genetic information in any familiar sense. Rather, the set of molecules and the reactions they undergo and catalyse constitutes the 'genome' of the system. The stable dynamical behaviour of this self-reproducing, coupled system of reactions constitutes the fundamental heritability it exhibits. The capacity to incorporate novel molecular species, and perhaps eliminate older molecular forms, constitutes the capacity for heritable variation. Darwin then tells us that such systems will evolve by natural selection.

If these considerations are correct, then, I submit, Schrödinger's suggested requirement for a large aperiodic solid as a stable carrier of heritable information is not necessary to the emergence of life or its evolution. Order from order in this sense, in short, may not be necessary.

Finally, I would like to mention briefly some experimental approaches to these questions. The fundamental question is this: if a sufficiently great diversity of polymers together with the small molecules of which they are composed, plus some other sources of chemical energy, were gathered in a sufficiently small volume under the appropriate conditions, would collectively autocatalytic sets emerge? These new experimental approaches rely on new molecular genetic possibilities. It is now feasible to clone essentially random DNA, RNA, and peptide sequences, creating extremely high diversities of these biopolymers (Ballivet & Kauffman 1985; Devlin *et al.*, 1990; Ellington & Szostak, 1990). At present, libraries with diversities of up to trillions of sequences are under exploration. Thus, for the first time it becomes possible to consider creating reaction systems with this high molecular diversity confined to small volumes such that rapid interactions can occur. For example, such polymers might be confined not only to continuous flow stirred reactors, but to vesicles such as liposomes, micelles, and other structures, which provide surfaces and a boundary between an internal and external environment. Given von Kiederowski's collectively autocatalytic sets, designed with his chemist's intelligence (personal communication, 1994), we know that such collectively autocatalytic sets of molecules can be constructed *de novo*. The phase transition theory I have outlined suggests that sufficiently complex systems of catalytic polymers should 'crystallize' connected, collectively autocatalytic webs of reactions as an emergent, spontaneous property, without the chemist's intelligent design of the web structure.

The emergence of collective autocatalysis depends upon how easy it is to generate polymers able to function both as substrates and as catalysts. This should not be extremely difficult. The existence of catalytic antibodies suggests that finding an antibody capable of catalysing an arbitrary reaction might require searching about a million to a billion antibody molecules. The binding site in the V region of an antibody molecule is nearly a set of several random peptides, corresponding to the complement determining regions, held in place by the remaining framework. Thus, libraries of more or less random peptides or polypeptides are reasonable candidates to serve as both substrates and catalysts. Indeed, recent work in collaboration with my graduate student Thomas LaBean and Tauseef Butt has shown that such random polypeptides tend readily to fold into a molten globule state, many of which show cooperative unfolding and refolding in graded denaturing conditions, suggesting modest folding capacities may be common in amino acid sequences (LaBean *et al.*, 1990, 1994). The results also suggest that random polypeptides may well exhibit a variety of liganding and catalytic functions. Earlier evidence in support of this rests on the display of random hexapeptides on the outer coat of filamentous phage. The probability of finding a peptide able to bind a monoclonal antibody molecule raised against another peptide is about one in a million (Devlin *et al.*, 1990; Scott & Smith, 1990; Cwirla *et al.*, 1990). Since binding a ligand and binding the transition state of a reaction are similar, these results, coupled with the success in finding catalytic antibodies, suggest that random peptides may rather readily catalyse reactions among peptides or other polymers. Random RNA sequences are interesting candidates as well. Recent results searching random RNA libraries for sequences which bind an arbitrary ligand suggest the probability of success is about one in a billion (Ellington & Szostak, 1990). Even more recent results seeking RNA sequences able to catalyse a reaction suggest a probability of about one in a trillion. It may prove easier to find random peptide sequences able to catalyse an arbitrary reaction. These results, coupled with rough estimates of the number of reactions such systems afford, suggest that diversities of perhaps 100 000 to 1 000 000 polymer sequences of length 100 might achieve collective autocatalysis.

THE SOURCES OF DYNAMICAL ORDER

If Schrödinger's suggestion is not necessary to the emergence of life, then might it at least be the case that the aperiodic solid of DNA is either necessary or sufficient to ensure heritable variation? The answer, I shall try to show in more detail than sketched above is, 'no'. The microcode enabled by the large aperiodic solid is clearly not sufficient to ensure order. The genome specifies a vast parallel processing network of activities. The dynamical behaviour of such a network could be catastrophically chaotic, disallowing any selectable heritability to the wildly varying behaviours of the encoded system. Encoding in a stable structure such as DNA cannot, by itself, ensure that the system encoded behaves with sufficient order for selectable heritable variation. Further, I shall suggest that encoding in a large stable aperiodic solid such as DNA is not necessary to achieve the stable dynamical behaviour requisite for selectable heritable variation of either primitive collectively autocatalytic sets, or of more advanced organisms. What may be required instead is that the system be a certain kind of open thermodynamic system capable of exhibiting powerful convergence in its state space towards small, stable dynamical attractors. The open system, in another view, must be able to discard information, or degrees of freedom, rapidly enough to offset thermal and other fluctuations.

I now summarize briefly the behaviour of random Boolean networks. These networks were first introduced as models of the genomic regulatory systems coordinating the activities of the thousands of genes and their products within each cell of a developing organism (Kauffman, 1969). Random Boolean networks are examples of highly disordered, massively parallel processing, non-equilibrium systems, and have become the subject of increased interest among physicists, mathematicians, and others (Kauffman, 1984, 1986, 1993; Derrida & Pommeau, 1986; Derrida & Weisbuch, 1986; Stauffer, 1987).

Random Boolean networks are open thermodynamic systems, displaced from equilibrium by an exogenous source of energy. The networks are systems of binary, on–off variables, each of which is governed by a logical switching rule, called a Boolean function. Boolean functions are named in honour of George Boole, a British logician who invented mathematical logic in the last century. Thus, one binary variable might receive inputs from two others, and be active at the next moment only if both input one AND input two were active the moment before. This is the logical or Boolean AND function. Alternatively, a binary variable with two inputs might be active at the next moment if either one OR the other or both of the two inputs is active at the current moment. This is the Boolean OR function.

Figure 3a–c shows a small Boolean network with three variables, each receiving inputs from the other two. One variable is assigned the AND function, the other two are assigned the OR function. In the simplest class

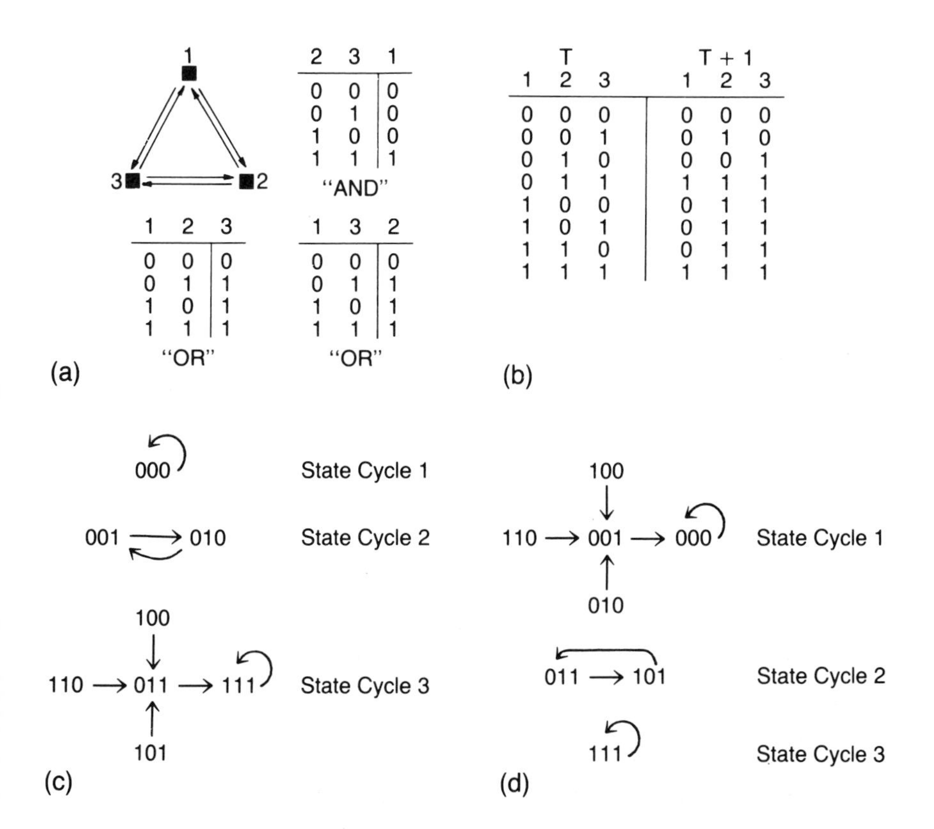

Figure 3. (a) The wiring diagram in a Boolean network with three binary elements, each an input to the other two. (b) The Boolean rules of (a) rewritten to show, for all $2^3 = 8$ states at time T, the activity assumed by each element at the next moment, $T + 1$. Read from left to right, this figure shows the successor state for each state. (c) The state transition graph, or behaviour field, of the autonomous Boolean network of (a) and (b) obtained by showing state transitions to successor states connected by arrows. (d) Effects of mutating the rule of element 2 from OR to AND.

of Boolean networks, time is synchronous. At each clocked moment, each element assesses the activities of its inputs, looks up the proper response in its Boolean function, and assumes the specified value. Also, in the simplest case, the network receives no inputs from outside. Its behaviour is fully autonomous.

While Figure 3a shows the wiring diagram of interconnections among the three variables, and the Boolean logical rule governing each, Figure 3b shows the same information in a different format. Define a state of the entire network as the current activities of all the binary variables. Thus, if there are N binary variables, then the number of states is just 2^N. In the present case, with 3 variables, there are just 8 states. The set of possible states of the network constitutes its state space. The left column of Figure 3b shows these 8 states. The right column shows the response, at the next moment, of each variable for every possible combination of activities of its inputs. However, another way of reading Figure 3b is to realize that the rows of the right half of the figure correspond to the next activities of all 3 variables. Hence read from left to right, Figure 3b specifies for each state of the entire network what its successor state will be.

Figure 3c shows the integrated dynamical behaviour of the entire network. This figure is derived from Figure 3b by drawing an arrow from each state to its unique successor state. Since each state has a unique successor state, the system will follow a trajectory of states in its state space. Since there is a finite number of states, the system must eventually re-enter a state previously encountered. But then, since each state has a unique successor state, the system will thereafter cycle repeatedly around a recurrent cycle of states, called a state cycle.

Many important properties of Boolean networks concern the state cycles of the system, and the character of trajectories flowing to such state cycles. Among these properties, the first is the length of a state cycle, which might be a single state which maps to itself forming a steady state, or the state cycle might orbit through all the states of the system. The state cycle length gives information about the recurrence time of patterns of activities in the network. Any Boolean network must have at least one state cycle, but may have more than one. The network in Figure 3c has three state cycles. Each state lies on a trajectory which flows to or is part of exactly one state cycle. Thus, state cycles drain a volume of state space, called a basin of attraction. The state cycle itself is called an attractor. A rough analogy sets state cycles equal to lakes and a basin of attraction equal to the drainage basin flowing into any single lake.

Examination of Figure 3c shows that trajectories converge. Trajectories converge onto one another either before reaching a state cycle, or, of course, converge when they reach the state cycle. This means that these systems throw away information. Once two trajectories have converged, the system no longer has any information to discriminate the pathway by which it arrived at its current state. Consequently, the higher the convergence in state space, the more information the system is discarding. We shall see in a moment that this erasure of the past is essential to the emergence of order in these massive networks.

Another property of interest concerns the stability of state cycles to minimal perturbations, transiently reversing the activity of any single variable. Examination of Figure 3c shows that the first state cycle is unstable to all such perturbations. Any such perturbation leaves the system in the basin of attraction of a different attractor to which the system then flows. In contrast, the third state cycle is stable to any minimal perturbation. Each such perturbation leaves the system in the same basin of attraction, to which the system returns after the perturbation.

THE CHAOTIC, ORDERED, AND COMPLEX REGIMES

After nearly three decades of study, it has become clear that large Boolean networks generically behave in one of three regimes, a chaotic regime, an ordered regime, and a complex regime in the vicinity of the transition between order and chaos. Of the three, perhaps the spontaneous emergence of an ordered regime coordinating the activities of thousands of binary variables is the most stunning for our current purposes. Such spontaneous collective order, I believe, may be one of the deepest sources of order in the biological world.

I shall describe the chaotic regime, then the ordered regime, and end with the complex regime.

Before proceeding, it is important to characterize the kind of questions that are being posed. I am concerned to understand the typical, or generic, properties of large Boolean networks in different classes of networks. In concrete terms, I shall be concerned with networks with a large number of binary variables, N. I shall consider networks classified by the number of inputs per variable, K. And I shall consider networks with specific biases on the set of possible Boolean functions of K inputs. We shall see that, if K is low, or if certain biases are utilized, then even vast Boolean networks, linking the activities of thousands of variables, will lie in the ordered regime. Thus, control of a few simple construction parameters suffice to ensure that typical members of the class, or ensemble, exhibit order. The evolutionary implication is immediate: achieving coordinated behaviour of very large numbers of linked variables can be achieved by tuning very simple general parameters of the overall system. Large scale dynamical order is far more readily available than we have supposed.

The aim to study the generic properties of classes, or ensembles, of networks demands that random members of that ensemble be sampled for investigation. Analysis of many such random samples then leads to an understanding of the typical behaviours of members of each ensemble. Thus, we shall be considering randomly constructed Boolean networks. Once constructed, the wiring diagram and logic of the network is fixed.

We consider first the limiting case where $K = N$. Here each binary variable receives inputs from itself and all other binary variables. There is, consequently, only a single possible wiring diagram. However, such systems can be sampled at random from the ensemble of possible $K = N$ networks by assigning to each variable a random Boolean function on its N inputs. Such a random function assigns, at random, a 1 or a 0 response to each input configuration. Since this is true for each of the N variables, a random $K = N$ network assigns a successor state at random from among the 2^N states, to each state. Thus, $K = N$ networks are random mappings of the 2^N integers into themselves.

The following properties obtain in $K = N$ networks. First, the expected median state cycle length is the square root of the number of states. Pause to think of the consequences. A small network with 200 variables would then have state cycles of length 2^{100}. This is approximately 10^{30} states. If it required a mere microsecond for the system to pass from state to state, it would require some billions of times the history of the universe since the big bang 14 billion years ago to orbit the state cycle.

The long state cycles in $K = N$ networks allow me to make a critical point about Schrödinger's argument. Think of the human genome. Each cell in a human body encodes some 100 000 genes. As we all know, genes regulate one another's activities via a web of molecular interactions. Transcription is regulated by sequences of DNA such as *cis* acting promoters, TATA boxes, enhancers, and so forth. The activities of *cis* acting sites, in turn, are controlled by transacting factors, often proteins encoded by other genes, which diffuse in the nucleus or cell, bind to the *cis* acting site, and regulate its behaviour. Beyond the genome, translation is regulated by a network of signals, as are the activities of a host of enzymes, whose phosphorylation states govern catalytic and binding activities. In turn, the phosphorylation state is controlled by other enzymes, kinases, and phosphatases which are themselves phosphorylated and dephosphorylated. The genome and its direct and indirect product, in short, constitute an intricate web of molecular interactions. The coordinated behaviour of this system controls cell behaviour and ontogeny.

Suppose the genome specified regulatory networks which were similar to a $K = N$ network. The time scale to turn a gene on or off is on the order of a minute to perhaps ten minutes. Let us retain the idealization that genes and other molecular components of the genomic regulatory system are binary variables. A genome with 100 000 genes, harbouring the complexity of the human genome, is capable of a mind boggling diversity of patterns of gene expression: $2^{100\,000}$. The expected state cycle attractors of such a system would be a 'mere' $2^{50\,000}$ or $10^{15\,000}$. To sketch the scale, recall that a tiny model genome with only 200 binary variables would require billions of times the age of the universe to traverse its orbit; $10^{15\,000}$ is not a number whose meaning we can even roughly fathom. But no organism could be based on state cycles of such unimaginably vast periods.

In short, were the human genome, duly encoded by an aperiodic solid called DNA, to specify a $K = N$ genomic regulatory system, the order enshrined in the aperiodic solid would beget dynamic behaviour of no possible biological relevance. Selection due to heritable variations requires a repeated phenotype upon which to operate. A genomic system whose gene activity patterns were a succession of randomly chosen states that only repeated in $10^{15\,000}$ steps could not exhibit such a repeated phenotype upon which selection could usefully operate.

$K = N$ networks have state cycles whose expected length scales exponentially with the size of the system. I shall use this scaling to denote one aspect of the chaotic behaviour of such networks.

But there is another sense of chaos, closer to the familiar one, which $K = N$ networks exhibit. Such networks show overwhelming sensitivity to initial conditions. Tiny changes in initial conditions lead to massive changes in the subsequent dynamics. The successor state to each state is randomly chosen among the possible states. Consider two initial states which differ in the activity of only one of the N binary variables. The states (000000) and (000001) are an example. The Hamming distance between two binary states is the number of bits which are different. Here, the Hamming distance is 1. If the Hamming distance is divided by the total number of binary variables, 6 in this example, the fraction of sites which are different, here 1/6, is a normalized Hamming distance. Consider our two initial states differing by a single bit.

Their successor states are randomly chosen among the possible states of the network. Hence the expected Hamming distance between the successor states is just half the number of binary variables. The normalized distance jumps from $1/N$ to $1/2$ in a single state transition. In short, $K = N$ networks show the maximum possible sensitivity to initial conditions.

Continuing my disagreement, if such it be, with the thrust of Schrödinger's book, were the human genome a $K = N$ network, not only would its attractor orbits be hyperastronomically long, but the smallest perturbations would lead to catastrophic alterations in the dynamical behaviour of the system. Once we have the counterexample of the ordered regime, it will become intuitively obvious that $K = N$ systems, deep in the chaotic regime, cannot be the way the genomic regulatory system is organized. Most importantly, selection operates on heritable variations. In $K = N$ networks, minor alterations in network structure or logic also wreak havoc with all the trajectories and attractors of the system. For example, deletion of a single gene eliminates half the state space, namely those in which that gene is active. This results in a massive reorganization of the flow in state space. Biologists wonder about possible evolutionary pathways via 'hopeful monsters'. Such pathways are highly improbable. $K = N$ networks could only evolve by impossibly hopeful monsters. In short, $K = N$ networks supply virtually no useful heritable variation upon which selection can act.

A word is needed about the term 'chaos'. Its definition is clear and established for systems of a few continuous differential equations. Such low dimensional systems fall onto 'strange attractors' where local flow is divergent but remains on the attractor. It is not clear, at present, what the relation is between such low dimensional chaos in continuous systems and the high dimensional chaos I describe here. Both behaviours are well established, however. By high dimensional chaos, I shall mean systems with a large number of variables in which the lengths of orbits scale exponentially with the number of variables, and which show sensitivity to intitial conditions in the sense defined above.

Order for free: Despite the fact that Boolean networks may harbour thousands of binary variables, unexpected and profound order can emerge spontaneously. I believe this order is so powerful that it may account for much of the dynamical order in organisms. Order emerges if very simple parameters of such networks are constrained in simple ways. The simplest parameter to control is K, the number of inputs per variable. If $K = 2$ or less, typical networks lie in the ordered regime.

Imagine a network with 100 000 binary variables. Each has been assigned at random $K = 2$ inputs. The wiring diagram is a mad scramble of interconnections with no discernible logic, indeed with no logic whatsoever. Each binary variable is assigned at random one of the 16 possible Boolean functions of two variables, AND, OR, IF, Exclusive OR, etc. The logic of the network itself is, therefore, entirely random. Yet order crystallizes.

The expected length of a state cycle in such networks is not the square root of the number of states, but on the order of the square root of the number of variables. Thus, a system of the complexity of the human genome, with some 100 000 genes and $2^{100\,000}$ states, will meekly settle down and cycle among a mere 317 states. And 317 is an infinitesimal subset of the set of $2^{100\,000}$ possible states. The relative localization in state space is on the order of $2^{-99\,998}$.

Boolean networks are open thermodynamic systems. In the simplest case, they can be constructed of real logic gates, powered by an exogenous electrical source. Yet this class of open thermodynamic systems shows massive convergence in state space. This convergence shows up in two ways. Overall, such systems exhibit a profound lack of sensitivity to initial conditions. The first signature of convergence is that most single bit perturbations leave the system on trajectories which later converge. Such convergence occurs even before the system has reached attractors. Secondly, perturbations from an attractor typically leave the system in a state which flows back to the same attractor. The attractors, in biological terms, spontaneously exhibit homeostasis. Both signatures of convergence are important. The stability of attractors implies repeatable behaviour in the presence of noise. But convergence of flow even before attractors are reached implies that systems in the ordered regime can react to similar environments in 'the same' way, even if ongoing perturbations by the environmental inputs persistently prevent the system from attaining an attractor. Convergence along trajectories should allow such systems to adapt successfully to a noisy environment.

Such homeostasis, reflecting convergence in state space, stand in sharp contrast to the perfect conservation of phase volume in closed, equilibrium thermodynamic systems. Recall that Liouville's theory ensures such conservation, which, in turn, reflects the reversibility of closed systems and the failure to throw away information into a heat bath. This conservation then underlies the capacity to predict probabilities of macrostates by the fractional number of microstates contributing to each macrostate.

The more important implication of conservation of phase volume in equilibrium systems is the following: Schrödinger correctly drew our attention to the fact that fluctuations in any classical system vary inversely with the square root of the number of events considered. When the system is an equilibrium system, these fluctuations have a given amplitude. However, if we consider an open thermodynamic system with massive convergence in state space, then that convergence tends to offset the fluctuations. The convergence tends to squeeze the system towards attractors, while the fluctuations tend to drive the system randomly in its space of possibilities. But if the convergence is powerful enough, it can confine the noise induced wandering to remain in the infinitesimal vicinity of the attractors of the system. Thus, we arrive at a critical conclusion. The noise induced fluctuations due to small numbers of molecules which concerned Schrödinger can, in principle, be offset by the convergent flow towards attractors if that flow is sufficiently convergent. Homeostasis can overcome thermalization.

But this conclusion is at the heart of the issue I raise with Schrödinger. For I want to suggest the possibility that the use by organisms of an aperiodic solid as the stable carrier of genetic information is not sufficient to ensure order. The encoded system might be chaotic. Nor is the aperiodic solid necessary. Rather, the convergent flow of systems in the ordered regime is both necessary and sufficient for the order required.

LATTICE BOOLEAN NETWORKS AND THE EDGE OF CHAOS

A simple modification of random Boolean networks helps understand the ordered, chaotic, and complex regimes. Instead of thinking of a random wiring diagram, consider instead a square lattice, where each site has inputs from its four neighbours. Endow each binary valued site with a random Boolean function on its four inputs. Start the system in a randomly chosen initial state, and allow the lattice to evolve forward in time. At each time step, any variable may change value from 1 to 0 or 0 to 1. If so, colour that variable green. If the variable does not change value, but remains 1 or remains 0, colour it red. Green means the variable is 'unfrozen' or 'moving'; red means the variable has stopped moving and is 'frozen'.

Random lattice networks with four inputs per variable lie in the chaotic regime. As one watches the lattice, most of the sites remain green; a few become red. More precisely a green unfrozen 'sea' spans or percolates across the lattice, leaving behind isolated red frozen islands.

I now introduce a simple bias among all possible Boolean functions. Any such function supplies an output value, 1 or 0, for each combination of values of its K inputs. The set of output values might be nearly half 1 and half 0 values, or might be biased towards all 1 values or all 0 values. Let P measure this bias. P is the fraction of input combination which give rise to the more frequent value, whether it be 1 or 0. For example, for the AND function, three of the four input configurations yield a 0 response. Only if both inputs are 1 is the regulated variable 1 at the next moment. P is therefore 0.75 in this case. Thus, P is a number between 0.5 and 1.0.

Derrida and Weisbuch (1987) showed that a Boolean lattice will lie in the ordered regime if the Boolean functions assigned to its sites are randomly chosen, with the constraint that the P value at each site is closer to 1.0 than a critical value. For a square lattice the critical value, P_c, is 0.72.

Consider a similar 'movie' of a network in the ordered regime, in which moving sites are again coloured green, and frozen sites are coloured red. If P is greater than P_c, then at first most sites are green. Soon, however, an increasing number of sites freeze into their dominant value, 1 or 0, and are coloured red. A vast red frozen sea spans or percolates across the lattice, leaving behind isolated green islands of unfrozen variables which continue to twinkle on and off in complex patterns. The percolation of a red frozen sea leaving isolated unfrozen green islands is characteristic of the ordered regime.

A phase transition occurs in such lattice Boolean networks as P is tuned from above to below P_c. As the phase transition is approached from above, the green unfrozen islands grow larger and eventually fuse with one another to form a percolating green unfrozen sea. The phase transition occurs just at this fusion.

With this image in mind, it becomes useful to define 'damage'. Damage is the propagating disturbances in the network after transiently reversing the activity of a site. To study this, it suffices to make two identical copies of the network, and initiate them in two states which differ in the activity of a single variable. Watch the two copies, and colour purple any site in the perturbed copy which is ever in a different activity value than its unperturbed copy. Then a purple stain spreading outward from the perturbed site demarcates the spreading damage from that site.

In the chaotic regime, let a site in the green percolating unfrozen sea be damaged. Then, generically, a purple stain spreads throughout most of the green sea. Indeed, the expected size of the damaged volume scales with the size of the total lattice system (Stauffer, 1987). Conversely, damage a site in the ordered regime. If that site lies in the red frozen structure, virtually no damage spreads outward. If the site lies in one of the green unfrozen islands, damage may spread throughout much of that island, but will not invade the red frozen structure. In short, the red frozen structure blocks the propagation of damage, and thus provides much of the homeostatic stability of the system.

At the phase transition, the size of distribution of damage avalanches is expected to be a power law, with many small avalanches and few large ones. The phase transition is the complex regime. In addition to the characteristic size distribution of damage avalanches, the mean convergence along trajectories which are near Hamming neighbours is zero. Thus, in the chaotic regime, initial states which are near Hamming neighbours tend, on average, to diverge from one another as each flows along its own trajectory. This is the 'sensitivity to initial conditions' of which I have spoken. In the ordered regime, nearby states tend to converge towards one another, often flowing into the same trajectory before reaching a common attractor. At the edge of chaos phase transition, on average, nearby states neither converge nor diverge.

It is an attractive hypothesis that complex adaptive systems may evolve to the complex regime at the edge of chaos. The properties of the edge of chaos regime have suggested to a number of workers (Langton 1986, 1992; Packard, 1988; Kauffman, 1993) that the phase transition, or edge of chaos, regime may be well suited to complex computations. On the face of it, the idea is attractive. Suppose one wished such a system to coordinate complex temporal behaviour of widely separated sites. Deep in the ordered regime, the green islands which might carry out a sequence of changing activities are isolated from one another. No coordination among them can occur. Deep in the chaotic regime, coordination will tend to be disrupted by any perturbations which unleash large avalanches of change. Thus, it is very plausible that near the phase transition, perhaps in the ordered regime, the capacity to coordinate complex behaviours might be optimized.

It would be fascinating if this hypothesis were true. We would begin to have a general theory about the internal structure and logic of complex, parallel processing, adaptive systems. According to this theory, selective adaptation for the very capacity to coordinate complex behaviour should lead adaptive systems to evolve to the phase transition itself, or to its vicinity.

Tentative evidence is beginning to support the hypothesis that complex systems may often evolve, not precisely to the edge of chaos, but to the ordered regime near the edge of chaos. In order to test this, my colleagues and I at the Santa Fe Institute are allowing Boolean networks to coevolve with one another to 'play' a variety of games. In all cases, the games involve sensing the activities of the other networks' elements and mounting an appropriate response to some of a network's own output variables. The coevolution of these networks allows them to alter K, P, and other parameters in order to optimize success at each game by natural selection. In brief summary, such networks do improve at the set of games we have asked them to perform. As always, such an evolutionary search takes place in the presence of mutational random drift processes which tend to disperse an adapting population across the space of possibilities it is exploring. Despite this drift tendency, there is a strong tendency to evolve toward a position within the ordered regime not too far from the transition to chaos. In short, tentative evidence supports the hypothesis that a large variety of parallel processing systems will evolve to the ordered regime near the phase transition in order to coordinate complex tasks.

Future work in this arena will examine the question central to that which I raise with Schrödinger. Two sources of 'noise' might occur in such game-playing Boolean networks. The first derives from inputs arriving from other networks. These exogenous inputs drive each system from its current trajectory, and hence perturb its flow towards attractors. The second is thermal noise within any one network. That internal noise will tend to perturb the behaviour of the system. In order to compensate and achieve coordination, such systems would be expected to shift deeper into the ordered regime. There the convergence in state space is stronger, and hence provides a more powerful buffer against exogenous noise. Thus, we can ask: how much convergence is required to offset a given amount of internal noise?

The same issue arises in any system whose dynamic behaviour is controlled by small numbers of copies of each kind of molecule. This occurs in contemporary cells, where the number of regulatory proteins and other molecules per cell are often in the range of a single copy. The same issue arises in the collectively autocatalytic molecular systems which I suspect may have formed at the dawn of life. How much convergence in state space offsets fluctuations due to the use of small numbers of molecules in a dynamical system, and how does the requisite convergence scale with the decrease in the number of copies of each kind of molecule in the model system? With respect to collectively autocatalytic sets of molecules, presumably some sufficiently high convergence in state space will buffer such systems from fluctuations due to the potentially small numbers of copies of each kind of molecule in the collectively reproducing metabolism. If so, the stable structure of large aperiodic solids is neither necessary nor sufficient to the order required for the emergence of life or those heritable variations upon which selection can successfully act.

ORDER AND ONTOGENY

We have seen that even random Boolean networks can spontaneously exhibit an unexpected and high degree of order. It would simply be foolish to ignore the possibility that such spontaneous order may play a role in the emergence and maintenance of order in ontogeny. While the evidence is still tentative, I believe the hypothesis finds considerable support. I shall briefly describe the evidence that genomic regulatory networks actually lie in the ordered regime, perhaps not too far from the edge of chaos. First, if one examines known regulated genes in viruses, bacteria, and eukaryotes, most are directly controlled by few molecular inputs, typically from 0 to perhaps 8. It is fascinating that, in the on–off Boolean idealization, almost all known regulated genes are governed by a biased subset of the possible Boolean functions which I long ago named 'canalizing' functions (Kauffman 1971, 1993; Kauffman & Harris, 1994). Here, at least one molecular input has one value, 1 or 0, which alone suffices to guarantee that the regulated locus assumes a specific output state, either 1 or 0. Thus, the OR function of four inputs is canalizing, since the first input, if active, guarantees that the regulated element be active regardless of the activities of the other three inputs. Boolean networks with more than $K = 2$ inputs per element, but confined largely to canalizing functions, generically lie in the ordered regime (Kauffman, 1993). I have, for some years, interpreted the attractors of a genetic network, the state cycles, as the cell types in the repertoire of the genomic system. Then the lengths of state cycles predict that cell types should be very restricted recurrent patterns of gene expression, and also predict that cells should cycle in hundreds to thousands of minutes. Further, the number of attractors scales as the square root of the number of variables. If an attractor is a cell type, we are led to predict that the number of cell types in an organism should scale as about the square root of the number of its genes. This appears to be qualitatively correct. Humans, with about 100 000 genes, would be predicted to have about 317 cell types. In fact humans are said to have 256 cell types (Alberts *et al.*, 1983) and the number of cell types appears to scale according to a relationship that lies between a linear and a square root function of genetic complexity (Kauffman, 1993). The model predicts other features such as the homeostatic stability of cell types. The frozen red component predicts, correctly, that about 70% of the genes should be in the same fixed states of activity on all cell types in the organism. Further, the sizes of green islands predict reasonably well the differences in gene activity patterns in different cell types of one organism. The size distribution of avalanches seems likely to predict the distribution of cascading alterations in gene activities after perturbing the activities of single randomly chosen genes. Finally, in the ordered regime, perturbations can only drive the system from one attractor to a few others. If attractors are cell types, this property predicts that ontogeny must be organized around branching pathways of differentiation. No cell type should, nor indeed can, differentiate directly to all cell types. Here is a property which presumably has remained true of all multicelled organisms since the Cambrian period or before.

A brief presentation of these ideas is all space allows. However, a fair summary at present is that genomic regulatory systems may well be parallel processing systems lying in the ordered regime. If so, then the characteristic convergence in state space of such systems is a major source of their dynamical order.

But there is a more dramatic implication of the self-organization I discuss here. Since Darwin we have come to believe that selection is the sole source of order in biology. Organisms, we have come to believe, are tinkered together contraptions, *ad hoc* marriages of design principles, chance, and necessity. I think this view is inadequate. Darwin did not know the power of self-organization. Indeed, we hardly glimpse that power ourselves. Such self-organization, from the origin of life to its coherent dynamics, must play an essential role in this history of life, indeed, I would argue, in any history of life. But Darwin was also correct. Natural selection is always acting. Thus, we must rethink evolutionary theory. The natural history of life is some form of marriage between self-organization and selection. We must see life anew, and fathom new laws for its unfolding.

SUMMARY

Schrödinger, writing before he had any right to have guessed so presciently, correctly foresaw that current life is based on the structure of large aperiodic solids. The stability of those solids, he foresaw, would provide the stable carrier material of genetic information. The microcode within such material would specify the organism. Quantum alterations in the material would be discrete, rare, and constitute mutations. He was correct about much of contemporary life.

But at a more fundamental level, was Schrödinger correct about life itself? Is the structural memory of the aperiodic solid necessary for all life? Surely, in the minimum sense that organic molecules with covalent bonds are small 'aperiodic solids', Schrödinger's argument has general merit. At least for carbon based life, one needs bonds of sufficient strength to be stable in a given environment. But it is the behaviours of collections of those molecules which constitute life on Earth and at least we may presume underlies many potential forms of life anywhere in the universe. Living organisms are, in fact, collectively autocatalytic molecular systems. New evidence and theory, adduced above, suggest that the emergence of self-reproducing molecular systems does not require large aperiodic solids. Limited evolution of such systems does not, in principle, require large aperiodic solids. Nor is dynamical order and heritable variation assured by an aperiodic solid which encodes the structure and some of the interactions of a large number of other molecules. Rather, heritable variation in self-reproducing chemical systems upon which natural selection can plausibly act requires dynamical stability. This, in turn, can be achieved by open thermodynamic systems which converge sufficiently in their state spaces to offset the fluctuations which derive from the fact that only small numbers of molecules are involved.

It is no criticism of Schrödinger that he did not consider the self-organized behaviours of open thermodynamic systems. Study of such systems had hardly begun fifty years ago and is not much advanced today. Indeed, all we can genuinely say at present is that the kinds of self-organization which we begin to glimpse in such open thermodynamic systems may be changing our view of the origin and evolution of life. It is enough that Schrödinger foresaw so much. We can only wish his wisdom were alive today to help further his and our story.

REFERENCES

Alberts, B., Bray, D., Lewis, J., Raff, M., Roberts K. & Watson, J. D. (1983). *Molecular Biology of the Cell*. New York: Garland.

Bagley, R. J. (1991) The functional self-organization of autocatalytic networks in a model of the evolution of biogenesis. Ph.D. thesis, University of California, San Diego.

Bagley, R. J. *et al.* (1992). Evolution of a metabolism. In *Artificial Life II*. A Proceedings Volume in the Santa Fe Institute Studies in the Sciences of Complexity, vol. 10, eds. C. G. Langton, J. D. Farmer, S. Rasmussen & C. Taylor, pp. 141–158. Reading, Massachusetts: Addison-Wesley.

Ballivet, M. & Kauffman, S. A. (1985). Process for obtaining DNA, RNA, peptides, polypeptides or proteins by recombinant DNA techniques. International Patent Application, granted in France 1987, United Kingdom 1989, Germany 1990.

Cohen, J. E. (1988). Threshold phenomena in random structures. *Disc. Appl. Math.*, **19**, 113–118.

Cwirla, P., Peters, E. A., Barrett, R. W. & Dower, W. J. (1990). Peptides on phages A vast library of peptides for identifying ligands. *Proceedings of the National Academy of Sciences USA* **87**, 6378–6382.

Derrida, B. & Pommeau, Y. (1986). Random networks of automata: A simple annealed approximation. *Europhysics Letters* **1**, 45–49.

Derrida, B. & Weisbuch, G. (1987). Evolution of overlaps between configurations in random Boolean networks. *Journal de Physique* **47**, 1297–1303.

Devlin, J. J., Panganiban, L. C. & Devlin, P. A. (1990). Random peptide libraries: a source of specific protein binding molecules. *Science* **249**, 404–406.

Eigen, M. (1971). Self-organization of matter and the evolution of biological macromolecules. *Naturwissenschaften* **58**, 465–523.

Ellington, A. & Szostak, J. (1990). In vitro selection of RNA molecules that bind specific ligands. *Nature* **346**, 818–822.

Erdos, P. & Renyi, A. (1960). *On the Evolution of Random Graphs*. Institute of Mathematics, Hungarian Academy of Sciences, publication no. 5.

Farmer, J. D., Kauffman, S. A. & Packard, N. H. (1986). Autocatalytic replication of polymers. *Physica* **22D**, 50–67.

Kauffman, S. A. (1969). Metabolic stability and epigenesis in randomly connected nets. *Journal of Theoretical Biology* **22**, 437–467.

Kauffman, S. A. (1971). Cellular homeostasis, epigenesis and replication in randomly aggregated macromolecular systems. *Journal of Cybernetics* **1**, 71.

Kauffman, S. A. (1984). Emergent properties in random complex automata. *Physica* **10D**, 145–156.

Kauffman, S. A. (1986). Autocatalytic sets of proteins. *Journal of Theoretical Biology* **119**, 1–24.

Kauffman, S. A. (1993). *The Origins of Order: Self Organization and Selection in Evolution*. New York: Oxford University Press.

Kauffman, S. A. & Harris, S. (1994). Manuscript in preparation.

LaBean, T. *et al.* (1992) Design, expression and characterisation of random sequence polypeptides as fusions with ubiquitin. *FASEB Journal* 6A471.
LaBean, T. *et al.* (1994). Manuscript submitted.
Langton, C. (1986). Studying artificial life with cellular automata. *Physica* **22D**, 120–149.
Langton, C. (1992). Adaptation to the edge of chaos. In *Artificial Life II,* A Proceedings Volume in the Santa Fe Institute Studies in the Sciences of Complexity, vol. 10, eds. C. G. Langton, J. D. Farmer, S. Rasmussen & C. Taylor, pp. 11–92. Reading, MA: Addison-Wesley.
Orgel, L. (1987). Evolution of the genetic apparatus: a review. In *Cold Spring Harbor Symposium on Quantitative Biology*, vol. 52. New York, NY: Cold Spring Harbor Laboratory.
Packard, N. (1988). Dynamic patterns in complex systems. In *Complexity in Biologic Modeling*, eds. J. A. S. Kelso & M. Shlesinger, pp. 293–301. Singapore: World Scientific.
Rossler, O. (1971). A system-theoretic model of biogenesis. *A. Naturforsch.* **B266**. 741.
Schrödinger, E. (1944). *What is Life?* Reprinted (1967) with *Mind and Matter* and *Autobiographical Sketches*. Cambridge: Cambridge University Press.
Scott, J. K. & Smith, G. P. (1990) Searching for peptide ligands with an epitope library. *Science* **249**, 386.
Stauffer, D. (1987). Random Boolean networks: Analogy with percolation. *Philosophical Magazine B* **56**, 901–916.
von Kiederowski, G. (1986). A self-replicating hexadesoxynucleotide. *Angewandte Chemie International Edition in English* **25**, 932–935.

9

Why new physics is needed to understand the mind

ROGER PENROSE
Mathematical Institute, Oxford

WHY CONSCIOUS UNDERSTANDING IS NON-COMPUTATIONAL

There are many facets to human mentality. It may well be that some of these can be explained in terms of our present-day physical concepts (compare Schrödinger, 1958) and, moreover, are potentially amenable to computational simulation. The proponents of artificial intelligence (AI) would maintain that such a simulation is indeed possible – at least for a good many of those mental qualities that are basically involved in our intelligence. Furthermore, such a simulation could be put to use in enabling a robot to behave, in those particular respects, in the same kind of way as a human being might behave. The proponents of *strong* AI would maintain, moreover, that *every* mental quality can be emulated – and will eventually be superseded – by the actions of electronic computers. They would also maintain that such mere computational action must evoke the same kind of conscious experiences in a computer, or robot, as we experience ourselves.

On the other hand, there are many who would argue to the contrary: that there are aspects of our mentality that cannot be addressed merely in terms of computation. Human consciousness, on such a view, would be such a quality – so it is *not* simply a manifestation of computation. Indeed, I shall argue so myself; but more than this, I shall argue that those actions which our brains perform in accordance with conscious deliberations must be things that cannot even be *simulated* computationally – so certainly computation cannot of itself give rise to any kind of conscious experience.

In order that such arguments can be made precise, it is necessary to have a clear-cut notion of what is *meant* by a 'computation'. Indeed, there is a mathematically precise definition of computation. This can be given in terms of what are known as *Turing machine* actions. A Turing machine is a mathematically idealized computer – idealized so that it can run on indefinitely without wearing out or slowing down, so that it never makes mistakes, and, most importantly, so that it has an unlimited storage capacity. (Thus, we must imagine that there is always the possibility for more storage capacity to be added, if there were danger of it running out.) I do not propose to give a more precise definition of a Turing machine than this, since the notion of 'computer' is now somewhat familiar to us. (For more details, see, for example, Penrose, 1989.)

As for the notion of 'consciousness', I shall make no attempt to define it here. All that we shall need about this notion is that whatever consciousness is, it is something that is necessarily present when we *understand* – when, in particular, we understand a mathematical argument.

Why do I claim that the effects of conscious deliberation cannot even be *simulated* by computational procedures? My own reasons stem, most powerfully, from the famous theorem of Kurt Gödel (1931). Gödel's theorem has the clear implication that mathematical understanding cannot be reduced to a set of known and fully believed computational rules. It is possible to go further than this and argue that no knowable set of purely computational procedures could lead to a computer-controlled robot that possesses genuine mathematical understanding. Such procedures could include not only deliberate 'top-down' algorithmic instructions, but also some more loosely programmed 'bottom-up' learning mechanisms. It is not really appropriate for me to enter into details; the full arguments will be given elsewhere (Penrose, 1994).

It would be unreasonable to suppose that there is anything particularly special about mathematical understanding, as opposed to other kinds of human understanding, with regard to the issue of non-computability. Accordingly, the non-computability of our mathematical understanding presumably has the implication that *every* kind of human understanding is also achieved by non-computable means. Likewise, it seems to me to be unreasonable to suppose that the various other aspects of human consciousness can be explained computationally any more than understanding can. Finally, I believe that non-human animals – at least many different kinds of animals – also possess the quality of consciousness and, consequently, must also act according to non-computational rules.

THE TWO LEVELS OF PHYSICAL ACTION

For the purpose of the remainder of our discussion, let us indeed take it that our brains act non-computationally when we indulge in conscious thought processes. Let us also accept that the actions of our brains are entirely governed by those same physical laws that underlie the behaviour of inanimate matter. Then we are confronted by the requirement that there must be physical actions that are governed by physical laws, yet which *in principle* cannot be simulated entirely computationally. What actions might these be?

We must first try to see whether there is scope, within the physical laws that are presently understood, for an appropriate non-computational behaviour. If we find that these laws fail to offer the scope that we need, then we must look beyond these laws in order to find the necessary non-computational processes. We must also ask for a plausible place where such non-computational physics could provide a key input into the functioning of our brains.

What, then, is the picture that today's physicists provide for us – as to the precise way that the physical world is understood to act? They would maintain that at the most fundamental level, the laws of quantum mechanics must hold. According to the Schrödinger picture, the state of the world at any one moment would be described by a *quantum state* (often denoted by the letter ψ, or by $|\psi\rangle$ in Dirac's notation) which describes a weighted combination of all the alternative ways in which the system under consideration *might* behave. It is not a probability-weighted combination, because the weighting factors are *complex numbers* (i.e. numbers of the form $a+ib$, where $i^2=-1$, a and b being ordinary real numbers). Moreover, the time-evolution of the quantum state is governed by a clear-cut deterministic equation, called the *Schrödinger equation.* The Schrödinger equation is a *linear* equation (leaving these complex weighting factors unaltered) and, in any ordinary sense, it would certainly be considered as providing a *computable* evolution for the quantum state. Thus quantum theory, in this sense, does not give us anything essentially non-computable.

However, Schrödinger evolution alone does not provide a picture of the world that makes sense at the classical level of phenomena (as Schrödinger himself was careful to emphasize). The rules of quantum linear superposition seem to apply only to states that differ very little from each other. Two states that differ very greatly from each other – such as two discernibly different locations for a golf ball – do not appear to exist in linear superposition. A golf ball, for example, has one location or it has another. It does not find itself in two places at once. An electron or a neutron, on the other hand, *can* exist in a superposition of two completely different locations at once (with complex-number weighting factors), and many experiments have been performed to confirm this kind of thing.

Thus, it appears, we must consider that there are two different levels of physical phenomena. There is the 'small-scale' *quantum* level, at which particles, atoms, or even molecules, can exist in these strange complex-number-weighted quantum superpositions. Yet, on the other hand, there is the *classical* level, at which one thing can happen, or another thing can happen, but where we do not experience complex combinations of alternatives. In particular, a golf ball is a classical-level object.

Of course, classical-level objects such as golf balls are themselves built from quantum-level constituents like electrons and protons. How is it that there can be one set of rules for the constituents and another for the large-scale object itself? In fact, this is a delicate issue which is not fully resolved within today's physical pictures. I shall need to return to the matter shortly; but, for the moment, it is best that we simply take the view that there are indeed two distinct levels of physical behaviour, and that there are different laws governing each.

HOW ARE THESE LEVELS BRIDGED?

At the quantum level, the mathematical description of a physical system is provided by the quantum state $|\psi\rangle$, as referred to above – sometimes called the system's *wavefunction*. So long as the system remains at the quantum level this state evolves with time (in the Schrödinger picture), in accordance with the deterministic and computable Schrödinger equation. I shall denote this evolution by **U** (unitary evolution). At the entirely classical level, the laws that control physical objects are those of Newton (for the commonplace motions of ordinary objects), of Maxwell (for the behaviour of electromagnetic fields), and of Einstein (when velocities or gravitational potentials become large). For all these classical types of evolution I shall use the notion **C**. These laws are again of a deterministic nature, and seem to be basically *computable* also. (In my assertions that **U** and **C** are 'computable', I have glossed over the fact that both **U** and **C** operate with continuous rather than the discrete parameters that are relevant to Turing-computability. We may suppose that adequate discrete approximations to **U** and **C** can be employed for this purpose, although this is less clear cut in the case of **C**, which is frequently chaotic, than it is in the case of the linear **U**.)

But how, in standard physical theory, does one deal with processes that involve both levels at once? Suppose, for example, that a physical system is so delicately poised that the behaviour of one of its quantum-level constituents can trigger off a large-scale classical effect? This is just the kind of situation that is referred to, in quantum theory, as a 'quantum measurement', and a different type of description from that entailed by Schrödinger's equation is needed. This is referred to as *state-vector reduction* (or collapse of the wavefunction), and I shall denote it by **R**. The standard mathematical procedure for describing a 'measurement' in quantum mechanics involves an instantaneous 'jump' from one quantum state to another. It is in the 'jumping' that is involved in quantum theory's **R**-procedure that all the probabilities and uncertainties of evolution arise; for a system that remains entirely at the quantum level, its evolution is described by the entirely deterministic and computable procedure **U**.

In any particular quantum measurement, the different possible outcomes are determined by the specific nature of the measurement that is being performed. All that the theory tells us is that there are certain possibilities attached to these various outcomes, these probabilities being determined by the particular quantum state that is being subjected to measurement. The theory makes no definite statement as to *which* of the possible outcomes takes place (except in special situations where the probabilities are 1 or 0). With regard to the results of measurements, *probability assignments* are all that the theory comes up with. Within the limitations that are provided by these probabilities the behaviour of the system is entirely *random*, whenever a measurement is performed.

Thus, present-day physical theory tells us that the objects of this world behave, for the most part, in an entirely computable way, except that from time to time (i.e. when a 'measurement' – or something equivalent – takes place) there is an additional entirely random ingredient in the behaviour of the system. Without this random ingredient, any physical system's behaviour would be regarded as *computational*, in the sense that a Turing-machine simulation could be set up which approximates the behaviour of the system as closely as is desired. Thus, we may regard a general physical system – as far as our present-day descriptions go – as something which behaves as a Turing machine with a randomizer.

However, a 'randomizer', in this sense, does not *in effect* give us anything that is beyond ordinary Turing computability. In practice, effectively random behaviour can be achieved by means of what are called 'pseudo-random' procedures, which means computational procedures that behave as random ones, for all intents and purposes. What would normally be done would be to have some 'chaotic calculation' which, though entirely computational in nature, would depend very critically upon some starting parameter. One might choose, for example, the precise *time* (as measured by the computer's clock) for this starting parameter. The result of this calculation would be, in effect, completely random, although it would be the result of the action of a Turing machine. There would be no difference in practice between using a pseudo-random computation of this nature and a genuinely random one.

The picture of physical reality provided by something that could be accurately modelled by a Turing machine with a randomizer – or with a pseudo-random input – does not, however, provide us with the kind of *non-computability* that the 'Gödelian' arguments of Section 1 tell us are necessary for the actions of a conscious brain. But is the 'pure randomness' that standard theory gives us *really* what is going on in a physical system? The weakest point in our present-day physical picture, at least at the kind of level that might be relevant to brain action, indeed lies in the random **R** process. Perhaps the present-day use of an entirely random '**R**' is merely a stop-gap? In my own opinion, this is indeed the case, and some new physical insights and a *new* physical theory will be needed in order to bridge the gap between **U** and **C**. In fact there is an increasing acceptance among (still a minority of) physicists that something needs to be done.

THE GRW STATE-REDUCTION SCHEME

One of the most promising recent suggestions for a modification of the rules of quantum theory, in accordance with aspirations of this nature, is that put forward by Giancarlo Ghirardi, Alberto Rimini and Tullio Weber (GRW). In their original scheme (Ghirardi, Rimini & Weber, 1986), GRW proposed that although a particle's wavefunction would evolve precisely according to the Schrödinger equation **U** for *most* of the time, there would be a small probability that the wavefunction would suffer a 'hit', whereby the wavefunction would be multiplied by another function, with a Gaussian spatial dependence. There are two arbitrarily chosen parameters in this theory, one of which (call it λ) determines the width of the Gaussian function, and the other (call it τ) which determines the rate at which the hits are taken to occur. The location of the peak of the Gaussian function is taken to be random, but with a probability distribution governed by the squared modulus of the wavefunction at the time at which it suffers the 'hit'. In this way, an agreement is achieved with the standard 'squared-modulus rule' governing the probabilities of ordinary quantum theory.

In the original GRW proposal, the value of τ is chosen so that a single particle on its own would suffer a 'hit' about once every 10^8 years. Thus, there is no conflict, over ordinary periods of time, with the standard quantum-mechanical descriptions of individual particles (whence, for example, the neutron diffraction experiments of Zeilinger *et al.*, 1988 are consistent with GRW). However, for systems involving large numbers of particles, the phenomenon of *quantum entanglement* must be taken into account. I shall describe this important phenomenon shortly, but for the moment, we just take note of the fact that, in standard quantum theory, the wavefunction of a system involving many particles must refer to the system in its entirety, and there is not simply a separate wavefunction for each particle individually. Thus, for a classical-level object with a large number of particles (say a golf ball), as soon as *one* of the constituent particles suffers a hit, the *entire* wavefunction of the object would be reduced. In the case of a golf ball, for which the number of particles would be about 10^{25}, the state would be reduced in less than a nanosecond. Thus, a quantum state which consists of a superposition of a golf ball in one location and the same golf ball in another location would be reduced, in a timescale of less than a nanosecond, *either* into the quantum state in which the golf ball has one of these locations *or* into the quantum state in which it has the other location.

In this way, the GRW scheme resolves one of the most fundamental problems encountered by the standard version of quantum theory: the *paradox of Schrödinger's cat* (Schrödinger, 1935a). According to this paradox, a cat is placed in a quantum superposition of two states, in one of which the cat is alive, and in the other, it is dead. Standard quantum theory – taken to insist that the evolution of a quantum state takes place solely according to the **U**-process – would tell us that the superposition of dead and alive cats must persist, and cannot resolve itself into one or the other. However, in the GRW scheme, the cat's state would indeed resolve itself into one or the other – in a timescale of a good deal less than a nanosecond.

ENTANGLED STATES

An important feature of the above scheme is that it depends upon the fact that a quantum state involving many particles is likely to be what is called an *entangled state*. I shall illustrate this type of situation in terms of what are called EPR (Einstein–Podolsky–Rosen) phenomena. These also serve to emphasize the essentially *non-local* nature of quantum states in relation to the **R**-process.

Imagine an initial state of spin 0 which decays into two particles, each of spin 1/2, these travelling off in opposite directions. If some specific direction in space is chosen, and the spin of each of these two particles is measured in that direction, then it has to be the case that the *opposite* answer is obtained for each particle – because the combined state of spin is zero. This applies whatever particular direction happens to be chosen.

More complicated measurements can also be performed, in which a *different* spin direction is chosen for each of the two particles. The measurement, in each case, yields merely the answer 'yes' or 'no' (since a spin 1/2 particle carries just one bit of information with regard to its spin). But there are certain joint probabilities, determined by standard quantum theory, for the answers for the two particles to *agree*, or to *disagree* (in fact $1-\cos\theta:1+\cos\theta$, where θ is the angle between the angles chosen).

Now according to a famous theorem of John S. Bell (1964), there is actually no 'local' way of explaining the joint probabilities that describe the quantum-mechanical predictions for these pairs of measurements, where each particle is regarded as a separate entity on its own. One must consider that the two particles are, in some mysterious way, still 'connected' to each other right up until a measurement is made on one or other of the particles. In effect, we have the situation that as soon as such a measurement on *one* particle is made, then this measurement instantaneously causes the state of the *other* particle to be reduced also. The state of the pair of particles cannot be thought of as being given by one particular state for one particle together with another particular state for the other particle. The *pair* of particles has a quantum state – an *entangled* state – but neither particle, separately, has a state on its own.

The phenomenon of quantum entanglement was first described by Erwin Schrödinger (1935b), as a general feature of quantum systems. Bell's theorem opened the way to experimental verification of the effects of quantum entanglements over large distances. These effects were subsequently observed by a number of experimenters, the most impressive experiments of this nature being performed by Alain Aspect and his colleagues (1982), where entanglements over distances of some twelve metres were observed.

Any ordinary large-scale object, such as the cat of Schrödinger's thought experiment, as described above, would itself be an entangled system. The individual particles of the cat's body would not have states on their own, but would be part of an entangled state for the cat as a whole. This would indeed be an implication of the standard quantum-mechanical descriptions. Now, the GRW scheme uses just the same kind of quantum-mechanical descriptions. Quantum entanglements are indeed just as much a feature of that scheme as they are for standard quantum mechanics. In this way, as soon as one of the particles of Schrödinger's cat's body suffers a GRW 'hit', the state of the entire cat is reduced with it, so the cat's state becomes either 'dead' or 'alive', rather than a quantum superposition of both together.

ENTANGLEMENT WITH THE ENVIRONMENT

In fact, the state of the cat would not be isolated from that of its environment, and we would have to consider that the entanglements do not end with the cat, but extend into this environment as well. Moreover, there would be many more particles involved in this disturbed environment than there would in the cat's body itself. In the *standard* discussions of the measurement process, the role of the environment of a quantum system is indeed considered to be all important. As the standard argument goes, the detailed quantum information (in what are called 'phase relations') that distinguishes a quantum superposition from a probability-weighted combination will simply get lost in these entanglements with the environment. Thus, *in effect* – or what John Bell has referred to as FAPP ('for all practical purposes') – a quantum superposition will behave like a probability-weighted combination of alternatives, as soon as entanglements with the random environment become significant.

However, all that this standard argument really achieves is a *coexistence* (FAPP) between the quantum procedures **U** and **R**, rather than a *deduction* of **R** from **U** (which would, in any case be impossible, strictly speaking, if only because the **U**-procedure itself makes no mention of probabilities). We still need something more than merely deterministic Schrödinger evolution

U of a system, if we are to explain how physical objects actually behave (as Schrödinger himself was careful to emphasize). What schemes like that of GRW attempt to achieve is a picture in which the physically observed process **R** – or something very like it – becomes part of the actual physical evolution of a system.

In fact, in the GRW scheme, it would be in the environment of a system that the 'hits' would normally occur first, and the entanglements of this environment with the system would result in the reduction procedure **R** taking effect in the system itself. For example, a DNA molecule would be far too small for 'hits' to be significant in the individual nucleotides of the molecule itself. Without 'hits' in the entangled environment playing their additional role, there would be nothing to determine the DNA strand as having a particular succession of nucleotides rather than merely having a quantum superposition of a number of different ones!

A NON-COMPUTABLE GRAVITATIONAL REDUCTION SCHEME?

None of this says anything about a role for non-computability in physical action, whose necessity was strongly argued for in Section 1. I have so far merely pointed to an important gap in our physical understanding – namely at the quantum/classical-level borderline – and mentioned one particular proposal (the GRW scheme) which attempts to bridge this gap. I would argue that there are strong reasons to believe that the physics which must *actually* bridge this gap will result from the appropriate union between quantum theory and Einstein's general theory of relativity. It is generally accepted that this union would have to result in a change in Einstein's gravitational theory (at very tiny distance scales), but it is a less conventional viewpoint that the standard rules of (**U**-)quantum theory must also have to change when the appropriate union is found, so that the phenomenon of state-vector reduction **R** will turn out to be a quantum-gravitational phenomenon (see Komar, 1969; Károlyházy, 1974; Károlyházy *et al.*, 1986; Diósi, 1989; Ghirardi *et al.*, 1990; Penrose, 1989, 1993, 1994).

If one accepts that the missing theory that is needed in order to replace the stop-gap **R** must be a gravitational one, then one is led to certain 'order-of-magnitude' estimates of the levels and timescales at which **R** ought actually to take place. (There are also some indirect – and rather tentative –

indications that such a theory might well be of a non-computational nature; cf. Penrose, 1994.)

To get some understanding of the level at which such a theory ought to begin to have relevance, consider the situation where a solid lump of material is placed in a quantum linear superposition of two distinct locations. I shall suppose that such a superposition is like an unstable particle or nucleus, with a certain half-life, and two separate modes of decay. In one of these modes, the superposed state decays into that state in which the lump occupies one of the two locations under consideration, and in the other mode, it decays into the state in which the lump occupies the other location. To estimate the half-life of this decay, we consider the energy E that it would take to displace the two instances of the lump away from each other, starting from coincidence, to the separation that they have in the superposition under consideration, where we take into account only the effect that the *gravitational* field of one of the lumps would have on the other (compare also Diósi, 1989). Another way of saying the same thing, assuming that the lump moves rigidly, is that E is the gravitational self-energy of the *difference* between the Newtonian gravitational fields of the two instances of the lump (cf. Penrose, 1994). The half-life T for the decay of the superposed state into one state of location or the other is then of the order of

$$T = \hbar/E,$$

where $\hbar$ is Planck's constant divided by 2π.

Let us examine this criterion in certain simple situations. If the lump consisted of just a single nuclear particle, then (taking the particle's radius to be of the order of one fermi) we get a decay time of some 10^7 years – similar to that of the original GRW scheme. For a lump of water density, we find that if its radius were one micrometre, then T would be about one twentieth of a second. If its radius were 10^{-3}cm, then T would be less than a millionth of a second; for a radius of 10^{-5}cm, then a few hours.

However, as argued above, these would be the reduction times only if the lump could remain isolated from its environment. If significant amounts of environmental material become disturbed, then the reduction time could be much shorter.

The proposal here is that for this state-reduction scheme to exhibit significant non-computational features, it would be necessary that the reduction takes place in the *system itself* – I shall refer to this as *self-reduction* – rather than in the environment. The idea is that the environment is essentially random, so that any genuinely non-computational features would be masked by this randomness, so long as it is state-reduction in the system's environment that (because of entanglement effects) reduces the system's state. In any normal experimental situation, it would indeed be the environment that controls the reduction of the state, so we see nothing different from the normal random behaviour that is so effectively described by the standard quantum-mechanical procedure **R**. It would take a very carefully organized structure for sufficient quantum isolation to occur that self-reduction takes place – before the random environment takes over. Such organization would be required so that there are significant *non-computational* deviations from the normal random **R**-procedure, as would be required according to the considerations of Section 1. No physical experimental set-up has yet got close to obtaining the necessary isolation.

RELEVANCE TO CONSCIOUS BRAIN ACTION

This is not to say that Nature herself has found no way to achieve the necessary conditions. Indeed, the arguments of Section 1 must be telling us that somehow she *has* found such a way. The conventional picture of brain action is that it can be completely understood in terms of nerve signals and synaptic action. Nerve signals seem to disturb their environment far too much for anything close to the isolation needed for the criteria of Section 7 to be achievable. But what about synaptic action? The strengths of (at least some) synapses are subject to continual change. What is it that controls these changes? There seem to be various different possibilities and different proposals, but one important factor seems to be the activity of the *microtubules* in the cytoskeletons of neurons.

What are microtubules? They are tiny tubelike structures which inhabit eukaryotic cells generally and which play many different kinds of role within cells. For example, in a one-celled animal they seem to be important in controlling its locomotion – as is the case for the continual change of shape of an amoeba. They exist in individual neurons, and control the ('amoeba-like') way that they connect with other neurons. They extend along the lengths of axons (perhaps not all in one piece) and dendrites, in each case right up to the close vicinities of the synapses. They transport various molecules along their lengths – in particular, the neurotransmitter chemicals that are vital for propagating neuronal nerve signals across synapses.

Mictrotubules are composed of a peanut-shaped protein called *tubulin* (about 8 nm × 4 nm × 4 nm in dimension), where the tubulins are arranged in a slightly skew hexagonal lattice. Each tubulin is a 'dimer', consisting of two components called 'α-tubulin' and 'β-tubulin'. A tubulin dimer is capable of existing in (at least) two different states – called 'conformations' (apparently depending upon the location of an electron, centrally placed in a 'hydrophobic pocket' between the two components). It has been suggested by Hameroff and Watt (1982; cf. also Hameroff, 1987) that these conformations give individual microtubules computer-like properties, where the two conformations of the dimers behave like the states 'on' or 'off', coding the bits '1' or '0' in a computer. Complicated signals could propagate along the microtubules in the manner of a cellular automaton.

So far this merely gives the potentiality for *computer*-type actions with enormously greater potential power than would be the case if we take individual neurons to be the sole 'computational units'. (Tubulin conformations act about a million times faster than neuron signals, and there are about ten million tubulins per neuron.) However, according to the foregoing discussion, we need something more than this, namely some scope for a *non-computational* action – an action which could arise only if a large-scale quantum-coherent state can be maintained in an isolated environment sufficiently long that the quantum state (or, at least, parts of the quantum state) can *self-collapse*, rather than collapsing because of entanglement with the environment. Do microtubules provide a plausible place for this kind of physical action? I believe that there is a good prospect that this may be so. Recall that microtubules are *tubes*. There is scope for some kind of quantum oscillation taking place *within* the tubes (Hameroff, 1974; del Giudice *et al.*, 1983; Hameroff, 1987; Jibu *et al.*, 1994; cf. also Fröhlich, 1968), which could be weakly coupled to the conformational actions, taking place along the tubes, of the tubulin dimers. The quantum oscillations within the tubes would probably not involve a significant amount of mass movement, but situations could arise when the coupling with the tubulin conformations becomes sufficiently great that there is just enough mass movement for self-collapse to occur. The viewpoint being presented here is that there is a non-computability involved in this self-collapse, and that conscious events are in some way to be identified with this process.

Clearly there is a good deal of speculation involved in these suggestions, but it seems to me that *something* of this general nature is needed. A much more complete presentation of these arguments is to be found in Penrose (1994) and a further article in preparation by Hameroff and Penrose.

REFERENCES

Aspect, A., Grangier, P. & Roger, G. (1982). Experimental realization of Einstein–Podolsky–Rosen–Bohm *Gedankenexperiment*: a new violation of Bell's inequalities. *Physical Review Letters* **48**, 91–4.

Bell, J. S. (1964). On the Einstein–Podolsky–Rosen paradox. *Physics* **1**, 195–200. Reprinted (1983) in *Quantum Theory and Measurement*, eds. J. A. Wheeler and W. H. Zurek Princeton: Princeton University Press.

del Giudice, E., Doglia, S. & Milani, M. (1983). Self-focusing and ponderomotive forces of coherent electric waves – a mechanism for cytoskeleton formation and dynamics. In *Coherent Excitations in Biological Systems*, eds. H. Fröhlich & F. Kremer. Berlin: Springer.

Diósi, L. (1989). Models for universal reduction of macroscopic quantum fluctuations. *Physical Review A* **40**, 1165–1174.

Fröhlich, H. (1968). Long-range coherence and energy storage in biological systems. *International Journal of Quantum Chemistry* **II**, 641–649.

Ghirardi, G. C., Rimini, A. & Weber, T. (1986). Unified dynamics for microscopic and macroscopic systems. *Physical Review* **34**, 470.

Ghirardi, G. C., Grassi, R. & Rimini, A. (1990). Continuous-spontaneous-reduction model involving gravity. *Physical Review A* **42**, 1057–1064.

Gödel, K. (1931). Über formal unentscheidbare Sätze der Principia Mathematica und verwandter Systeme I. *Monatshefte für Mathematik und Physik* **38**, 173–198.

Hameroff, S. R. (1974). Chi: a neural hologram? *American Journal of Clinical Medicine* **2**(2), 163–170.

Hameroff, S. R. (1987) *Ultimate Computing. Biomolecular Consciousness and Nano-Technology*. Amsterdam: North Holland.

Hameroff, S. R. & Watt, R. C. (1982). Information processing in microtubules. *Journal of Theoretical Biology* **98**, 549–61.

Jibu, M., Hagan, S., Hameroff, S. R., Pribram, K. H. & Yasue, K. (1994). Quantum optical coherence in cytoskeletal microtubules: implications for brain function. *BioSystems* **32**, 195–209.

Károlyházy, F. (1974). Gravitation and quantum mechanics of macroscopic bodies. *Magyar Fizikai Polyoirat* **12**, 24.

Károlyházy, F., Frenkel, A. & Lukács, B. (1986). On the possible role of gravity on the reduction of the wave function. In *Quantum Concepts in Space and Time*, eds. R. Penrose & C. J. Isham. Oxford: Oxford University Press.

Komar, A. B. (1969). Qualitative features of quantized gravitation. *International Journal of Theoretical Physics* **2**, 157–160.

Penrose, R. (1989). *The Emperor's New Mind: Concerning Computers, Minds, and the Laws of Physics*. Oxford: Oxford University Press.

Penrose, R. (1993). Gravity and quantum mechanics, in *General Relativity and Gravitation 1992. Proceedings of the Thirteenth International Conference on General Relativity and Gravitation held at Cordoba, Argentina, 28 June–4 July 1992. Part 1: Plenary Lectures*, eds. R. J. Gleiser, C. N. Kozameh & O. M. Moreschi. Bristol: Institute of Physics Publishing.

Penrose, R. (1994). *Shadows of the Mind: An Approach to the Missing Science of Consciousness*. Oxford: Oxford University Press.
Schrödinger, E. (1935a). Die gegenwärtige Situation in der Quantenmechanik. *Naturwissenschaften* **23**, 807–812, 823–828, 844–849. (Translation by J. T. Trimmer (1980). *Proceedings of the American Philosophical Society* **124**, 323–38. Reprinted in *Quantum Theory and Measurement*, eds. J. A. Wheeler & W. H. Zurek. Princeton: Princeton University Press, 1983.)
Schrödinger, E. (1935b). Probability relations between separated systems. *Proceedings of the Cambridge Philosophical Society* **31**, 555–563.
Schrödinger, E. (1958). *Mind and Matter*. Reprinted (1967) in *What is Life?* with *Mind and Matter* and *Autobiographical Sketches*. Cambridge: Cambridge University Press.
Zeilinger, A., Gaehler, R., Schull, C. G. & Mampe, W. (1988). Single and double slit diffraction of neutrons. *Reviews in Modern Physics* **60**, 1067.

10

Do the laws of Nature evolve?

WALTER THIRRING
Institut für Theoretische Physik, Universität Wien

Many things in Nature which we thought to be eternal, like the fixed stars, atoms, or quantities such as mass, turned out to be only temporary forms. Now the only thing which is believed to be of eternal stature is the law of Nature. In a contribution to a symposium at the Pontifical Academy of Science on 'Understanding Reality: The Rôle of Culture and Science' I tried to explain why I do not think that this is necessarily so and that also the laws may evolve in the course of the history of the universe. Herewith I would like to submit this heresy to a wider scientific public, not as an eternal truth but as a possibility worthy of reflection and discussion.

Today physics knows the laws which describe the material world from the tiny to the vast. That there are no phenomena known which contradict these laws is small wonder since whenever this happened, physics had to adjust them to accommodate also the hitherto unexplained events. The surprising fact is that in this process of gradual extension the laws became both more general and more unified. For instance, quantum mechanics of atomic systems supersedes classical mechanics and incorporates it as a limiting case. Similarly, elementary particle physics contains atomic physics as a low energy limit. This leads people to think that at the top of this pyramid there is an *Urgleichung* (as it was termed by Heisenberg; now it is called a TOE = Theory of Everything) which contains everything.

However this may be, whether this TOE will ever be found or remain a mirage forever, it left in its wake this pyramid of laws referring to particular space and time scales, some with restricted, some with vast, domains of applicability. Also biologists talk about various levels of laws (Novikoff, 1945; Mayr, 1988; Weinberg, 1987) except that they turn the pyramid around. The more complex the system the higher it ranges for the biologists. To the extent that 'up' and 'down' reflects involuntarily a value judgement this shows a difference between physicist and biologist in their attitude towards complexity.

The claim 'Theory of Everything' has of course to be understood within the framework of our present thinking. To formulate the laws of physics it is advisable to use the language of quantum theory and to talk of observables and states. For instance for one particle its position x and its momentum p are the observables, whereas the states are given by the Schrödinger function, which gives a probability distribution for these quantities. The observables are an objective reality and develop deterministically. By this I mean that there is a one to one correspondence between these quantities at a time t, $(x(t), p(t))$ and the initial quantities (x, p). More precisely, the transformations $(x, p) \to (x(t), p(t))$ form a one-parameter group of automorphisms of the algebra generated by (x, p), t being the parameter. The state in which a particular system is reflects our subjective knowledge and in quantum mechanics it is never complete, which leads to a certain unpredictability. (I do not want to use the term causality, which has another philosophical connotation.) In spite of the deterministic time evolution not everything can be predicted with certainty because even in the present there is some uncertainty. For large systems this uncertainty becomes overwhelming since we can only measure a small fraction of all the observables. Of course, we are free to choose what we want to measure but in any case it will be only a small part. Mathematically this means that the state can only be determined within some weak neighbourhood.

Nevertheless there is a general belief that the time evolution dictated by the *Urgleichung* contains the dynamics of the whole universe and determines everything. I would like to replace this view by another one based on three theses.

(i) The laws of any lower level in the pyramid mentioned above are not completely determined by the laws of the upper level though they do not contradict them. However, what looks like a fundamental fact at some level may seem purely accidental when looked at from the upper level.

(ii) The laws of a lower level depend more on the circumstances they refer to than on the laws above. However, they may need the latter to resolve some internal ambiguities.

(iii) The hierarchy of laws has evolved together with the evolution of the universe. The newly created laws did not exist at the beginning as laws but only as possibilities.

I do not consider these propositions as revolutionary but as plausible conjectures inspired by our present knowledge. Far from being able to prove them mathematically I shall illustrate them by some examples. I am aware of the fact that some of them belong to speculative parts of physics and may remain to be only fiction. They should be taken only as models to illustrate my points.

(a) That we live in a world with three space dimensions and one time dimension is the basis of our theories and many people have amused themselves by exploring how strange life would be in worlds with different numbers of dimensions. Yet the current way of thinking suggests that at the beginning the world had far more dimensions and by some anisotropy only three dimensions have expanded enormously (Chodos & Detweiler, 1980). By now the others have collapsed and left their traces only in internal symmetries of elementary particles. This splitting in $4 + x$ dimensions is by no means engraved in the *Urgleichung*, which is perfectly symmetric in all of them. In these theories this particular splitting appears accidentally and is as unpredictable as the position of a drop in a condensation phenomenon. Such an unpredictability seems to contradict the deterministic time evolution. After all, I have to take only the present state, evolve it backwards in time and then I know exactly what is the initial state that leads to the present situation. However, as mentioned, in large systems one can only determine in which weak neighbourhood a state is and the contention is that any weak neighbourhood contains states which will develop in all conceivable ways.

(b) Though this internal space which curled up to 10^{-33} cm had no preferred direction its symmetry got broken by some phase transition and the fundamental interaction got split into the strong, electromagnetic, and weak forces (Barrow & Tipler, 1986; Weinberg, 1977). Why it happened just this way was certainly not determined by the original thermal equilibrium and the latter must have contained potentially all laws emerging from one kind of symmetry breaking or another. For a long time it was an obsession of many great physicists to find a theory which explains the numerical value of the strength of these interactions, such as the famous fine-structure constant $e^2/\hbar c = (137.0 \ldots)^{-1}$. So far these attempts have failed and in the present picture their values appear to be accidental.

(c) Many-body dynamics does not depend so much on the detailed form of the interaction between the particles as on a property called stability (Lieb,

1991; Thirring, 1990). It states that the potential energy per particle is bounded from below by an energy independent of the number of particles. If this is not satisfied, matter forms a hot cluster which eventually may disappear in a black hole. That ordinary matter is stable hinges crucially on the fact that the electron obeys Fermi statistics. If the π^- were lighter than the electron and therefore the lightest stable charged particle we would be in a peculiar situation. Hydrogen $= p\pi^-$ would still be stable since the proton is a fermion, but deuterium $= d\pi^-$ not. For N charged bosons the ground state energy goes approximately as $-N^{7/5}$ and thus one mole of deuterium would contain $10^{24\times 2/5} \approx 10^{9.6}$ more energy per mole than hydrogen. In this scenario all nuclei would turn into isotopes of even mass and matter would become a superdense plasma.

(d) The long time stability of larger structures such as the planetary system are governed by resonances (Siegel & Moser, 1971). If the revolution times of two planets resonate, the smaller one is kicked out of its orbit. The fate of our Earth is thus determined by the number theoretic properties of the ratio of the revolution times of other planets (mainly Jupiter) and ours. Whether the law of force is Newton's $1/r^2$ or something else is of little importance. Thus for the issue of how long our Earth will enjoy sunshine, number theory is more important than the field theory of gravitation.

This example also shows why one may have to appeal to the upper level to resolve ambiguities. The singularity of the $-1/r$-potential prevents us from predicting classically whether a head-on orbit is reflected by the singularity or goes right through. In quantum mechanics this singularity is no problem for the time evolution and its classical limit tells us that the former alternative is the correct one.

The list of such examples could be continued at will and I would like to make the following observation.

To accommodate seemingly contradicting facts physics had to widen its concepts and thereby lost predictive power. For instance, quantum mechanics describes the wave and corpuscular properties of particles at the expense of the uncertainty relations. The *Urgleichung* – if there is such a thing – must potentially contain all possible routes which the universe could have been taking and therefore all possible laws. Clearly, it must leave a lot of leeway. With such an equation physics would be in a situation similar to mathematics around 1930 when Gödel showed that mathematical structures may not be inconsistent but will contain true statements which are not deducible. Similarly, the *Urgleichung* will not contradict experience, otherwise it would be modified, but will be far from determining everything. As the universe evolved, the circumstances created their own laws.

One might cherish the feeling that the different levels discussed above are all only different realizations of a more fundamental principle. The difficulty of making this intuition more precise lies in a good definition of what is fundamental. For instance, one can argue that the fundamental principle in field theory (classical or quantal) is Lorentz invariance and Maxwell's or Yang–Mills equations are only special mechanisms to exhibit this principle. Yet it is not of global validity: in general relativity we learn that spaces with such a large group of isomorphisms are rather exceptional. Or even worse, as discussed under (a) the dimensionality and signature of space-time may be the result of a historical accident. Similarly the extensive nature of the energy going approximately as N may be considered a fundamental law. It is simple and of wide validity; it holds for all chemical elements. It forms the basis of an important science, namely thermodynamics. Yet it is not of global validity and is violated by gravitational interactions. Again, as discussed under (b) it may be the result of a historical accident: were there a charged boson lighter than the electron the fundamental law would be energy $\propto N^{7/5}$ rather than energy $\propto N$. In this sense laws which appear fundamental to us may not have existed in the beginning as laws but only as possibilities.

These views may shift the emphasis from what is important in science. In the prevailing picture the noblest goal in science has to be to find the TOE since everything else then means only the working out of special cases. If one believes that the few Greek letters in the *Urgleichung* do not say much but that the real physics consists of its mathematical consequences in a given situation then the various levels in the pyramid of physics stand in their own right. This does not mean that one is not obliged to deduce from the upper levels what is deducible but only with due modesty and without false pretences.

REFERENCES

Barrow, J. D. & Tipler, F. (1986). *The Anthropic Cosmological Principle*. Oxford: Clarendon Press.

Chodos, A. & Detweiler, S. (1980). Where has the fifth dimension gone? *Physical Review D*21, 2167–2170.
Lieb, E. H. (1991). The stability of matter. In *From Atoms to Stars. Selected Papers*. New York: Springer.
Mayr, E. (1988). The limits of reductionism. *Nature* **331**, 475.
Novikoff, A. B. (1945). The concept of integrable levels in biology. *Science* **101**, 209–215.
Siegel, C. L. & Moser, J. (1971). *Lectures in Celestial Mechanics*. New York: Springer.
Thirring, W. (1990). The stability of matter. *Foundations of Physics* **20**, 1103–1110.
Weinberg, S. (1977). *The First Three Minutes – A Modern View of the Origin of the Universe*. London: André Deutsch.
Weinberg, S. (1987). Newtonianism, reductionism and the art of congressional testimony. *Nature* **330**, 433–437.

11

New laws to be expected in the organism: synergetics of brain and behaviour

J. A. SCOTT KELSO[1] and HERMANN HAKEN[1,2]

[1]*Program in Complex Systems & Brain Sciences, Center for Complex Systems, Florida Atlantic University, Boca Raton, Florida*

[2]*Institute for Theoretical Physics & Synergetics, University of Stuttgart, Stuttgart*

Acknowledgements

Much of the work described in this article is supported by NIMH (Neurosciences Research Branch) Grant MH42900, BRS Grant RR07258, Office of Naval Research Contract N00014-92-J-1904 and NSF Grant DBS-9213995. We are very grateful to Tom Holroyd and Armin Fuchs for their help with the figures.

One can best appreciate, from a study of living things, how primitive physics still is.
(A. Einstein)

INTRODUCTION

The title of this article – at least the statement in front of the colon – is unashamedly stolen from Schrödinger's (1944) wonderful little book *What is Life?* The statement after the colon points to a source where these new laws may be found. Synergetics is a term coined by H. Haken (1969, 1977) to encapsulate a relatively new multidisciplinary field of research aimed at understanding how patterns form in open, nonequilibrium systems, i.e. systems that receive a continuous influx of energy and/or matter. Synergetics deals with how the (typically very many) individual parts of a system cooperate to create novel spatiotemporal or functional structures. In the last decade or so tremendous progress has been made in penetrating nature's ways of generating patterns in open physical, chemical and biochemical systems (e.g. Babloyantz, 1986; Bak, 1993; Bergé, Pomeau & Vidal, 1984; Collet & Eckmann, 1990; Ho, in press; Iberall & Soodak, 1987; Kuramoto, 1984; Nicolis & Prigogine, 1989, for reviews). In particular, synergetic construction principles have established the concepts of instability, order parameters, fluctuations and slaving as crucial to understanding and predicting the spontaneous (self-organized) formation of pattern in complex systems.

When Schrödinger openly suggested – probably to the horror of many, then and now – that understanding living systems might involve 'other laws' beyond the 'known laws of physics', the theoretical concepts of pattern formation and self-organization in open, nonequilibrium systems were virtually unheard of (murmurings, however, are contained in von Bertalanfy's early work as well as Schrödinger's somewhat unfortunate introduction of the term 'negative entropy'). Likewise, the mathematical tools of nonlinear dynamics had yet to bloom, in part because computation – the principal means of exploring nonlinear equations whose analytic solutions are not known – was practically nonexistent. Bemoaning their intellectual efforts, Schrödinger once said of his esteemed colleagues Dirac and Eddington, in a letter to Born: 'That is the thing beyond their linear thoughts. All is linear, linear . . . If everything were linear, nothing would influence nothing, said Einstein once to me. That is actually so.' (Moore, 1989, p. 381.) Schrödinger recognized that the enormously interesting and important structures studied by physicists that arise owing to disorder–order transitions (e.g. when matter changes its macroscopic structure as temperature is lowered) were completely irrelevant to the emergence of life processes. In physics, different aggregate states of matter – solid, liquid, gas – are called *phases* and the transitions between them are called *phase transitions*. When vapour changes to liquid and eventually to ice, this is an example of progressive change from disorder to order. It is immediately clear that life processes have nothing to do with this kind of phase transition and that entirely different principles having to do with *nonequilibrium* phase transitions are needed. The latter occur in systems that are pumped or energized from the outside (or, like living systems that possess a metabolism, from inside or outside). Without an exchange of energy, matter or information with their surroundings, such systems cannot maintain their structure or function.

In biology, at least so far, processes of self-organization in open systems have been given short shrift. Certainly, reaction–diffusion mechanisms of the Turing type have been mentioned in discussions of embryological development and the genesis of biological form – how a cell becomes a finger or a toe – but, for the most part, it is a brief and passing 'tip of the hat' (e.g. Wolpert, 1991). Certainly, most biologists admit that organisms belong to the general class of open systems. In *Of Molecules and Men*, Crick (1966) even remarks – in a single paragraph – that 'the organism must be an open system' (p. 9). This, he says, is the first minimum requirement for life. Not surprisingly, however, by far the most attention is devoted to the fact that organisms possess genetic material that allows them to reproduce and pass on 'copies' of themselves to their descendants. Darwinian selection does the rest. Yet, it has long been recognized that Darwinian selection *presupposes* the existence of self-sustaining structures like the gene; it does not explain how that particular configuration was selected from the primordial soup. Indeed, there is as yet no experimental demonstration of biological order without the aid of (already ordered!) biological precursors (Dyson, 1985).

In short, modern biology recognizes that organisms are organized things. Over the course of time, biologists have taken great pains to discredit the slightest hint that some immaterial vital force underlies biological organization (e.g. Mayr, 1988), an attitude shared by us. Even though biology should have had the strongest motivation to expose the limitations of 'ordinary physics and chemistry' and to search out and explicate Schrödinger's 'new laws' (of self-organization in open systems?), it did not . . . Instead, it took a different path (molecular biology) which, despite enormous successes, it seems to be paying for now (see Maddox, 1993). To make our point of view quite clear: we do not claim that the fundamental laws of physics (and thus chemistry) do not hold in biology; they, of course, do. But we do claim that their conceptual frame is too narrow. Rather we have to find new concepts that transcend the purely microscopic description of systems.

Although this chapter does not deal with molecular events, *per se*, it does suggest that nonlinear processes taking place far from equilibrium are rich enough to handle biological self-organization on several different scales. The aim is to show that the physical concepts of self-organized pattern formation (i.e. synergetics) already provide a foundation for understanding organisms and their relation to the environment. The chapter is organized as follows. In Section 2, some of the main concepts of Haken's synergetics are introduced in the context of a familiar example from physics. In Section 3, these ideas are applied to the problem of *coordination*, which it is argued is a (perhaps *the*) fundamental feature of living things. In such complex systems *relevant* degrees of freedom and their dynamics are often not known but have to be found. Synergetics provides a level-independent strategy and methods to elucidate the underlying (nonlinear) dynamics. Section 4 presents some new evidence demonstrating that the brain itself is fundamentally an active, self-organizing system subject to nonlinear dynamical laws. Theory and experiment converge on the notion that biological systems, including the brain, live near boundaries separating regular and irregular behaviour, surviving best, as it were, in the margins of instability. In a final section, some of the implications of these results are drawn for life itself. Schrödinger's prose, incidentally, is impossible to emulate. Like him, however, the goal here is to convey the essential ingredients of biological self-organization in a conceptual and nontechnical fashion, and with a minimum of equations.

HOW NATURE HANDLES COMPLEXITY

Any account of pattern formation in open, nonequilibrium systems has to handle (at least) two problems. The first concerns how patterns are constructed from a very large number of material components. The second is that often not just one pattern but *multiple* patterns are produced to accommodate environmental conditions. Biological structures, for example, are multifunctional: the same set of components may self-organize for different functions or different components may self-organize for the same function. Moreover, how a given pattern or structure persists under various environmental conditions (its *stability*) and how it adjusts to changing internal or external conditions (its *adaptability*) have to be accounted for. The processes that determine how a pattern is *selected* from the myriad of possibilities must also be accommodated by any putative law or principle of self-organization. As we shall see, such processes often involve *cooperation* and *competition*, and a subtle interplay between the two.

To explain the mechanisms underlying pattern formation, let us use the familiar example of a fluid heated from below and cooled from above. First, a word of caution. No one is saying that the brain, or living things in general, are simply fluids composed of homogeneous elements. Far from it. Rather the fluid is used here as an example that illustrates some of the ways nature handles complex nonequilibrium systems containing many degrees of freedom. In particular, it allows us to illustrate the key concepts of synergetics that will provide a foundation for understanding the emergence of biological order. Like all great physical experiments, the beauty of the fluid example is that even though it is performed in the laboratory, it provides a window onto the bigger picture. The experiment is called the Rayleigh–Bénard instability and it goes like this. Take a liquid, say a little cooking oil, put it in a pan and heat it from below. Microscopically, the fluid contains, say, 10^{20} molecules, each of which is subject to random, disordered motion (*very many microscopic elements*). If the temperature difference between the top and the bottom of the fluid is small there will be no large scale motion of the fluid. Heat is dissipated among the elements as a micromotion that we cannot see. Notice, even at this stage, that this is an *open system*, activated by a temperature gradient which is called a *control parameter* in the language of synergetics and dynamical systems. As this control parameter increases, an amazing event called an *instability* occurs. The liquid begins to move macroscopically in an orderly rolling motion. The system is no longer a haphazard collection of randomly moving molecules: billions of molecules cooperate to create macroscopic patterns evolving in space and time. The reason for the onset of rolling motion (convection) is that the cooler liquid at the top of the fluid layer is more dense and tends to fall, whereas the warmer and less dense fluid at the bottom tends to rise.

In synergetics, the amplitude of the movement of the rolls plays the role of an *order parameter* or *collective variable*: all parts of the fluid no longer behave independently but are 'sucked' into an ordered mode of coordination. In the vicinity of critical regions (i.e. near an instability) the system's macroscopic behaviour is dominated by just a few collective modes, the so-called order parameters, which are the only variables needed to describe the evolving pattern formation exhaustively. This compression of degrees of freedom (*df*) near critical points is referred to in the physics literature as the *slaving principle*, due to Haken (1977) who has given it an exact mathematical form for a large class of systems. For an excellent review of the slaving principle, see Wunderlin (1987). Examples include vortex formation in the Taylor–Couette system, the onset of coherent laser light, the formation of concentration patterns in certain chemical reactions such as the Belousev–Zhabotinski reaction, and the well-studied Turing instability which has served, with limited success, as a model of morphogenesis. In all these cases, the emergence of pattern and pattern switching arises solely as a result of the cooperative dynamics of the system with no specific ordering influence from the outside and no homunculus-like agent or *program* inside. The control parameter is *non-specific*, that is, it does not prescribe or contain the code for the emergent pattern which is said to be a product of *self-organization*. In self-organizing systems there is no *deus ex machina*, no ghost in the machine ordering the parts. No 'self', in fact. Later on, we shall discuss how *specific* parametic influences on biological processes may be incorporated into this picture.

A few further points. One concerns *circular causality*: the order parameter is created by the cooperation of the individual parts of the system. Conversely, the order parameter governs the behaviour of the individual parts. For example, in the laser, the stimulated emission of atoms generates the light field, which in turn acts as an order parameter specifying or – in Haken's words – 'slaving' the motion of the electrons in the atoms. The outcome is an enormous compression of information. Circular causality is typical of nonlinear processes in far from (thermal) equilibrium conditions. It contrasts with the linear causality that dominates most of biology and physiology, e.g. the old 'central dogma' that information flows in only one direction from DNA to RNA to protein. A second point concerns fluctuations and symmetry breaking. How, in our physical example, does the rolling motion of the fluid know in which direction to flow? The answer is chance itself: symmetry of left or right handed motion is broken by an accidental fluctuation or perturbation. Once the 'decision' is made it is quite final and cannot be reversed. All the elements have to obey it. This interplay between chance (stochastic processes) and choice determines the patterns that emerge. In biological self-organizing systems, fluctuations are always present probing the stability of existing states and allowing the system to discover new ones. A third point is that more and more complex patterns – an entire hierarchy of instabilities – may arise as the control parameter is further increased. New patterns are created again and again in ever increasing complexity. Sometimes a system can be driven so hard that it goes into a turbulent state. There are too many options for the components to adopt and behaviour never settles down.

In summary, synergetics deals typically with equations of the following form:

$$\dot{\boldsymbol{q}} = \boldsymbol{N}(\boldsymbol{q}, \text{parameters, noise}), \qquad (1)$$

where the dot denotes the derivative with respect to time, $\boldsymbol{q}$ is a potentially high-dimensional state vector specifying the state of the system, Eq. (1), and $\boldsymbol{N}$ is a nonlinear function of the state vector and may depend on a number of parameters (including time) as well as random forces acting on the system. In general, when parameters in Eq. (1) change continuously, the corresponding solutions of Eq. (1) also change continuously. However, when a continuous change in the control parameter crosses a critical value, the system's behaviour may change qualitatively or discontinuously. Such qualitative changes are associated with the spontaneous (self-organized) formation of patterns and always arise via an instability. Patterns emerging at *nonequilibrium phase transitions* (the term preferred by physicists because it includes the effects of fluctuations) or *bifurcations* (the mathematical term used in dynamic systems theory) are defined in terms of *attractors* of the collective variable or order parameter dynamics. (Discussion of these terms will be further elucidated in the next section within the context of biological coordination.) Attractive states of the collective variable dynamics exist because nonequilibrium systems are *dissipative*: many independent trajectories with different initial conditions converge in time to a certain limit set or attractor solution. Often stable fixed point, limit cycle and chaotic solutions – as well as various other transient, more complicated behaviours – are possible in the *same* system, depending on parameter values. Here, then, is one of nature's main themes for handling complex living things (Kelso, 1988): enormous *material complexity* is compressed near instabilities (as demonstrated by the slaving principle of synergetics) giving rise to lower-dimensional behaviour that is described by collective variables or order parameters. The resulting pattern dynamics is nonlinear, from which emerges rich *behavioural complexity*, including stochastic features and/or deterministic chaos. This scenario provides a conceptual and mathematical foundation for the disorder–order and order–order principles advocated by Schrödinger (1944) and adds the evolutionary order-to-chaos principle of open, dissipative systems. The latter, it turns out, are chock full of 'new physics'.

COORDINATION DYNAMICS OF LIVING THINGS

I do not see any way to avoid the problem of coordination and still understand the physical basis of life. (H. Pattee)

In spite of, or perhaps because of, the successes of modern molecular biology the great unresolved problem of all biology remains: how complex living things are coordinated in space and time. Neither classical nor quantum physics (notwithstanding the declarations of physicists such as Hawking, Penrose and Weinberg) provide any insight into functionally specific coordination. Although we claim to know all the laws of the behaviour of matter ('ordinary physics and chemistry') except under extreme conditions, such laws hardly tell us a single iota about how or why we walk down the street. As Howard Pattee (1976) remarked years ago, the riddle of life has been solved by molecular biology. But there is more to life than the chemistry of cellular reactions. The origin and nature of the *coordination* of these reactions remain obscure. Imagine, for the moment, a living system composed of individual components that ignored each other and did not interact with either themselves or their surroundings. Such a system would possess neither structure nor function. Regardless of the level of description one chooses to study (the personal choice of the scientist, given one believes, as we do, that there is no *ontological* priority of any single level of description over any other) the degrees of freedom are (at least transiently) *coupled* or functionally linked. In the case of the brain, for example, the individual nerve cells do not think, smell, act or remember. Instead, they appear to cooperate together in temporally coherent groups to generate what we call cognitive functions. The essential questions for understanding coordination in living things concern the form that the basic interaction takes, how it occurs and why it is the way it is.

Putative solutions to these questions lie, at least in primitive form, in what may be called *elementary coordination dynamics* (Kelso, 1990, 1994). By elementary, one means a simple mathematical formulation (but not so simple that the essence of the problem is lost) that nevertheless provides a foundation for understanding other issues, such as learning and adaptation to the environment and the relation of these processes to brain function. Needless to say, elementary coordination dynamics uses the concepts of self-organization and pattern formation (Haken, 1977) as part of a theoretically motivated experimental strategy and the tools and language of coupled nonlinear dynamics to express lawfully (in continuous or discrete form) how coordination patterns form and change.

How do we find basic laws of coordination? Put another way: how do we find relevant collective variables for complex systems and their dynamics on a chosen level of observation? In line with synergetics, phase transitions (or bifurcations) provide a special entry point for developing theoretical understanding of complex living things in which the *relevant* degrees of freedom are usually not known. The reason is that *qualitative* change allows the clear distinction of one pattern from another, thereby enabling the identification of collective variables for different patterns and the pattern dynamics (multistability, loss of stability, etc.). Near critical points, the essential processes governing a pattern's stability, flexibility and even its selection can be uncovered. Theoretically motivated measurements (fluctuations, relaxation times, dwell times near the critical point, and so forth; see below) are available to elucidate these processes and to allow tests of theoretical predictions (e.g. Schöner & Kelso, 1988a; Kelso, Ding & Schöner, 1992). *Control* parameter(s) that promote instabilities can be determined. Instabilities provide a generic mechanism for flexible change (switching without switches) among coordinative patterns, that is, for getting in and out of coherent states. Finally, different levels of description – coordinative and individual component levels – can be related through a study of (uncoupled) component dynamics and their nonlinear coupling.

Strangely enough, basic coordination laws first became accessible in human motor coordination (Haken, Kelso & Bunz, 1985; Schöner, Haken & Kelso, 1986) following the experimental discovery of spontaneous, involuntary changes in hand movement patterns (Kelso, 1981, 1984) analogous, perhaps, to the spatiotemporal reordering that occurs when an animal changes locomotory gait (see Shik, Severin & Orlovski, 1966). When human subjects are requested to rhythmically move their index fingers in an alternating, out-of-phase fashion and frequency of motion is systematically increased, a spontaneous transition into a symmetric, in-phase pattern is observed. No such transition back from in-phase to anti-phase is observed as frequency is reduced. Likewise, when the system is prepared in the in-phase pattern and frequency is increased, no shift to the anti-phase coordination pattern occurs.

This simple experimental example is illustrative of an elementary coordinative linkage in complex, biological systems. It contains *essentially nonlinear* features of self-organization, namely, multistability (two coordinative states coexist for the same parameter values), transitions from one ordered state to another and hysteresis, a primitive kind of memory.

The simplest coordination dynamics that captures all the experimental results is

$$\dot{\phi} = -a \sin \phi - 2b \sin 2\phi, \tag{2}$$

where ϕ is the relative phase between the two rhythmically interacting components and the ratio b/a is a control parameter corresponding to the cycle period, t, the reciprocal of frequency. There is a good reason to suppose that ϕ is the relevant order parameter of coordination. First, it captures the spatiotemporal ordering among the components. All other observables, as it were, are 'slaved' to the phase relation. Second, it changes much more slowly than the variables describing the behaviour of the individual components. Third, ϕ changes abruptly at the transition. The dynamics of Eq. (2) can be visualized as a particle moving in the landscape of a potential function, $V(\phi)$. Therefore, an equivalent formulation of Eq. (2) is

$$\dot{\phi} = -\frac{\partial V(\phi)}{\partial \phi} \quad \text{with} \quad V(\phi) = -a \cos \phi - b \cos 2\phi. \tag{3}$$

The potential landscape or 'attractor layout' for different movement rates, i.e. different ratios of b/a, is plotted in Figure 4 (top).

This so-called HKB dynamics, (2) and (3), accommodates the observed coordination facts. (1) It has two stable fixed point attractors corresponding to phase- and frequency-locked states at $\phi = 0$ (in-phase) and $\phi = \pm \pi$ rad (anti-phase). For low values of the ratio b/a, both coordination modes coexist – which one is observed depends on initial conditions – the essentially nonlinear property of *bistability*. (2) As the ratio b/a is decreased the fixed point at π loses stability, any small fluctuation kicking the system into the only remaining stable fixed point at $\phi = 0$. Beyond this *spontaneous phase transition* only the symmetrical pattern at $\phi = 0$ is stable. And (3) when the direction of the control parameter change is reversed, the coordination system remains in the in-phase attractor. This *hysteresis* is due to the fact that the fixed point at $\phi = 0$ is always stable.

The basic coordination dynamics, Eqs. (2) and (3), has been extended in numerous ways that can only briefly be mentioned here. Among these extensions are:

- the introduction of stochastic forces to Eqs. (2) and (3) led to predictions of *critical slowing down* and *critical fluctuations* near the instability (Haken *et al.*, 1985; Schöner *et al.*, 1986). These predictions are easy to intuit from Figure 4 (top). As the minimum at $\phi = \pi$ becomes shallower and shallower, the system takes longer and longer to recover from any small perturbation. Thus, the relaxation time is predicted to increase as the instability is approached because the restoring force (the gradient of the potential) becomes smaller (critical slowing down). Likewise, the variability of ϕ is expected to increase (critical fluctuations) owing to the flattening of the potential near the transition point. Both predictions have been confirmed in a wide variety of experimental systems (e.g. Buchanan & Kelso, 1993; Kelso & Scholz, 1985; Kelso, Scholz & Schöner, 1986; Scholz, Kelso & Schöner, 1987; Schmidt, Carello & Turvey, 1990; Wimmers, Beek & van Wieringen, 1992).

- The influence of *specific* parametric influences has been incorporated in Eq. (2), e.g. when a particular pattern is specified by the environment, learning and intention (e.g. Kelso, Scholz & Schöner, 1988; Schöner & Kelso, 1988b; Zanone & Kelso, 1992). A benefit of knowing Eq. (2), in which coordination patterns form and change owing to *non-specific* parametric influences (i.e. the control parameter, b/a, simply moves the system through its collective states, but does not prescribe them), is that it allows *specific* parameters from various sources to be expressed dynamically (i.e. as 'forcings' defined in exactly the same language as the order parameter(s)). A conceptual advantage is that the duality between (specific) information and (non-specific, intrinsic) dynamics is removed. Information, according to this scheme, is only meaningful and specific to a living system to the extent that it contributes to the order parameter dynamics, attracting them to the required coordination pattern. Whether such a theoretical perspective can contribute to the 'real problem' of life (Rosen, 1991), namely, how to move the holonomic (symbolic, rate-independent) order characteristic of a DNA or RNA sequence into nonholonomic (rate-dependent, 'behaving') order manifest by the phenotype is an open question. Rather, the present analysis suggests a reformulation of the problem. Here, self-organized coordination laws like Eq. (2) are, at their very roots, *informational* structures. The identified order parameter, ϕ, captures the coherent relations among different kinds of things. Unlike 'ordinary physics' the order parameter for biological coordination is context dependent and intrinsically meaningful to system functioning. What, one asks, could be more meaningful to an organism than information that specifies the coordinative relations between its parts or between itself and the environment?
- The inclusion of a symmetry breaking term in Eq. (2) in order to accommodate situations in which the component units are not identical, e.g. where the uncoupled components exhibit different eigenfrequencies. Notice that Eq. (2) is a *symmetric* coordination law: the system is 2π periodic and identical under left–right reflection ($\phi \rightarrow -\phi$). Nature, of course, thrives on broken symmetry, the sources and consequences of which are

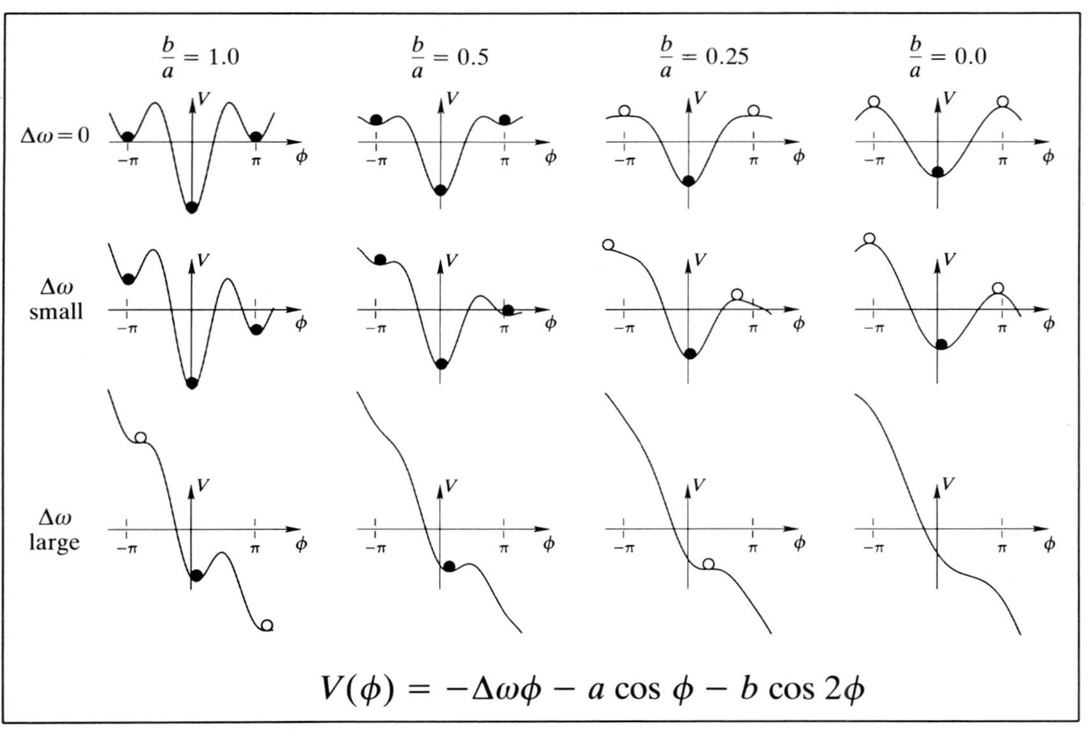

Figure 4. The HKB potential as a function of the ratio b/a for different values of $\Delta\omega$. Solid circles indicate stable fixed point attractors; open circles are unstable fixed points. *Top*: $\Delta\omega = 0$: the potential is symmetric, with minima initially located at $\phi = 0$ and $\phi = \pm\pi$ (see text). *Middle*: $\Delta\omega$ small: The potential is asymmetric, with minima slightly shifted. *Bottom*: $\Delta\omega$ large: Only the shifted minimum near $\phi = 0$ is initially stable, then it too disappears. Notice how 'remnants' of the previously stable fixed points remain at certain parameter values of b/a.

manifold in living things. The coordination dynamics, Eq. (2), can be readily extended to incorporate symmetry breaking by adding a constant term, $\Delta\omega$, equivalent to the frequency difference among (uncoupled) components (Kelso, DelColle & Schöner, 1990). Ignoring stochastic forces, the dynamics now become

$$\begin{aligned} \dot{\phi} &= \Delta\omega - a \sin \phi - 2b \sin 2\phi, \text{ and} \\ V(\phi) &= -\Delta\omega\phi - a \cos \phi - b \cos 2\phi \end{aligned} \tag{4}$$

for the equation of motion and the potential respectively. Figure 4 (middle, bottom) shows the evolution of the attractor layout for different values of $\Delta\omega$. This extension predicts two important consequences of symmetry breaking. First, for small values of $\Delta\omega$ it predicts that the minima of the potential are no longer at $\phi = 0$ and $\phi = \pi$ but are systematically shifted. Second, for large enough values of $\Delta\omega$ there are no longer local minima in the attractor layout – the stable fixed points disappear – and the relative phase undergoes drift. Again, both predictions have been observed experimentally (Kelso *et al.*, 1990; Kelso & Jeka, 1992; Schmidt, Shaw & Turvey, 1993; see also contributions in Swinnen *et al.*, 1994).

Notice in Figure 4 (bottom) that even though there is no longer strict coordination, 'remnants' or 'ghosts' of fully coordinated states remain, e.g. near $\phi = 0$. This is termed *intermittency* and represents one of the generic processes found in low-dimensional systems near tangent or saddle-node bifurcations. As a result of broken symmetry in the coordination dynamics, the system – instead of being absolutely coordinated – exhibits a partial or relative coordination between its components. Relative coordination, as von Holst (1939) remarked years ago, is 'a kind of neural cooperation that renders visible the operative forces of the central nervous system that would otherwise remain invisible'. The effect arises owing to *competing* tendencies for full coordination (phase- and frequency-locking) on the one hand and the tendency of the individual components to express their intrinsic spatial and temporal variation on the other. One can readily see this from the coordination dynamics Eq. (4), in which the ratio b/a represents the relative importance of the intrinsic phase attractive states at 0 and π, and $\Delta\omega$ corresponds to frequency differences between the components. The identification of this more variable, plastic and fluid form of relative coordination with the dynamical mechanism of intermittency (Kelso,

DeGuzman & Holroyd, 1991) is consistent with the emerging view that biological systems tend to live near boundaries between regular and irregular behaviours (Kauffman, 1993). By occupying the strategic, intermittent region near the boundaries of mode-locked states, living things (and the brain itself, see below) are afforded the necessary mix of stability (of the hyperbolic, not asymptotic kind) and the ability to flexibly switch among 'metastable' coordinated states.

- It is probably obvious that Eqs. (2) and (4) can readily be elaborated for the coordination of multiple, anatomically different components (e.g. Collins & Stewart, 1993; Schöner, Jiang & Kelso, 1990; Jeka, Kelso & Kiemel, 1993). Experimental research has identified these individual components as nonlinear oscillators which – as archetypes of time-dependent behaviour – are essential ingredients of the dynamics of non-monotonic evolution, whether regular or irregular (Bergé *et al.*, 1984). Recently, Jirsa, Friedrich, Haken & Kelso (1994) have postulated that the original HKB coupling

$$K_{12} = (\dot{X}_1 - \dot{X}_2)\{\alpha + \beta(X_1 - X_2)^2\}. \tag{5}$$

in which α and β are coupling parameters and X_1 and X_2 correspond to self-sustaining nonlinear oscillators, may be a fundamental biophysical coupling. The reason is that Eq. (5) provides the simplest way to couple components so as to guarantee properties that are quite critical to living things: multistability, flexibility and transitions among co-ordinated states. Another reason, of course, is that the basic self-organized coordination dynamics, Eqs. (2) and (4), can be derived using Eq. (5).

In summary, Eqs. (5) and (4) represent elementary forms of coupling and coordination, respectively. The basic coordination dynamics contain (a) no coordination; (b) absolute coordination (when two or more components synchronize at the same frequency and maintain a fixed relation; and (c) relative coordination (the *tendency* toward phase attraction even when the component frequencies are not the same). All these different forms of self-organization have an explanation, namely, they are patterns that emerge in different parameter régimes of the identified coordination dynamics. At the heart of such dynamics lies a spatiotemporal symmetry which, when broken, generates an event structure for living things that includes pattern formation, pattern switching and intermittency. The dynamics Eqs. (2), (3), and (4) have been shown experimentally to express coordination between (a) components of an organism; (b) organisms themselves; and (c) organisms and their environment (see Kelso, 1994, for a review) and provide a foundation for further theoretical and experimental developments, one of which is treated next.

SELF-ORGANIZATION IN THE BRAIN

Is the brain itself a self-organized, pattern forming system? Specifically, do phase transitions exist in the brain and, if so, what form do they take? How is it possible to capture the immense patterned complexity in space and time of Sherrington's 'enchanted loom'? At least three things are needed to answer these questions: an appropriate set of theoretical concepts and corresponding methodological strategies; a technology that affords analysis of the global dynamics of the brain; and a clean experimental paradigm that prunes away complications but retains essential aspects. In this section, recent work is summarized (see Kelso *et al.*, 1991, 1992; Fuchs, Kelso & Haken, 1992; Fuchs & Kelso, 1993 for details) that attempts to incorporate all these features.

The experiment involves transitions in sensorimotor coordination in a paradigm introduced by Kelso, DelColle and Schöner (1990). A subject (see also Wallenstein, Bressler, Fuchs & Kelso, 1993) is exposed to periodic acoustic stimuli and instructed to press a button in between two consecutive tones, i.e. to syncopate with the stimulus. The stimulus frequency starts at 1 Hz and is increased in 8 steps by 0.25 Hz after every ten tones. At a certain critical frequency the subject is no longer able to syncopate and switches spontaneously to a coordination pattern that is now synchronized with the stimulus. During these runs, brain activity is recorded using a 37-SQUID array located over left parieto-temporal cortex as shown in Figures 5a, b and c. SQUIDs (superconducting quantum interference devices) allow access to the spatiotemporal patterning of magnetic fields generated by intracellular dendritic current flow in the brain. Because the skull and scalp are transparent to magnetic fields generated inside the brain and because the sensor array is large enough to cover a substantial portion of human neocortex, this new research tool opens a (non-invasive) window into the brain's spatiotemporal organization and its relation to real-time behaviour.

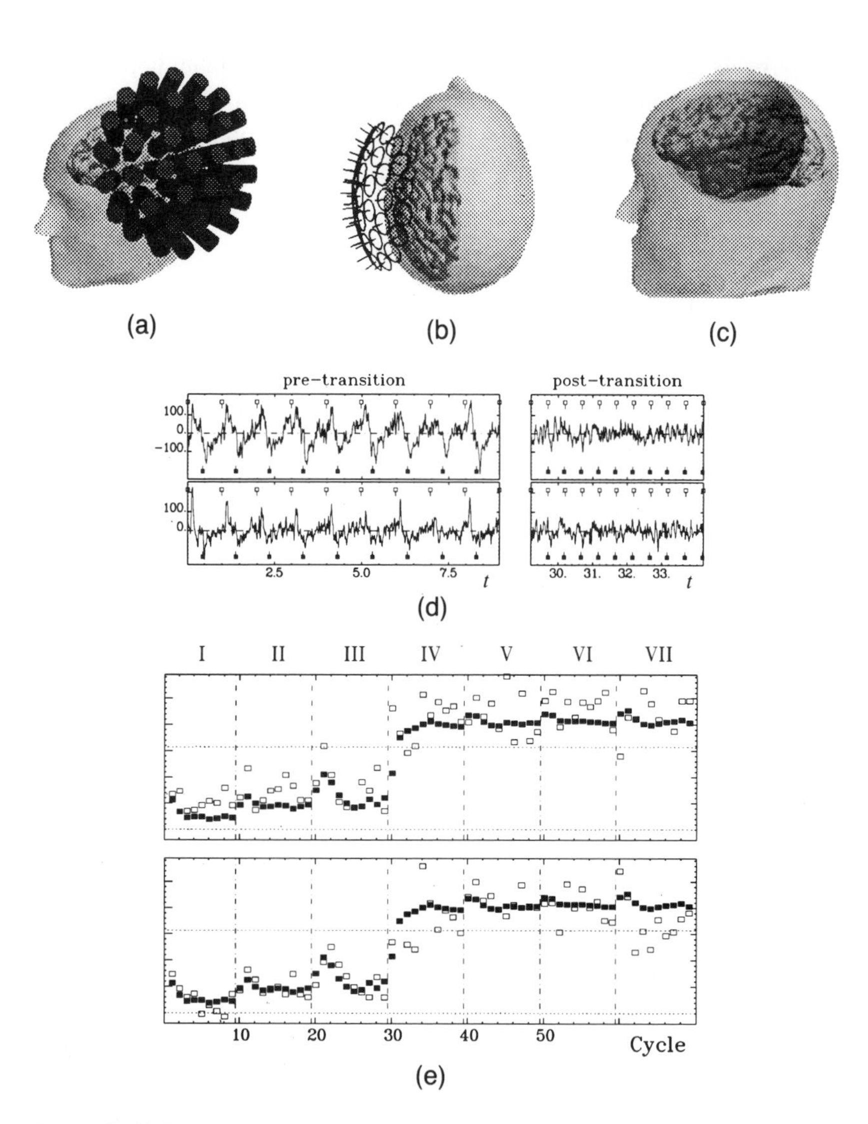

Figure 5. (a) Reconstruction of the subject's head and location of SQUID sensors. (b) Construction of the cortex model, using magnetic resonance imaging. Slices were taken in a coronal plane with a spacing of 3.5 mm. The location and orientation of each SQUID sensor is superimposed. (c) Example of magnetic field activity detected by SQUIDs and displayed on the head-cortex model. (d) Time series from two single sensors before and after the transition. (e) Superimposed relative phase (y-axis) calculated at the stimulus frequency for each cycle for the behaviour over time (solid squares) and two of the sensors (open squares). See text for details.

Figure 5d shows the averaged data from two SQUID sensors before and after the behavioural transition from syncopation to synchronization. Open squares mark the point in time where the stimulus occurred; solid squares correspond to the (right finger) button press. Before the transition, the stimulus and the response are anti-phase. After the transition the subject's responses are nearly in-phase with the stimulus. The neural activity of the brain shows a strong periodicity during this perception–action task, especially in the pre-transition region. After the transition, the amplitude drops (even though the movements are more rapid) and the signals look noisier. This result is paradoxical, but extremely interesting. On the one hand, behavioural synchronization is more stable than syncopation. On the other, brain activity during synchronization is less coherent than syncopation, as can be easily seen in Figure 5d. The difficulty of the task conditions appears to determine the coherence of the signal.

A most remarkable result is shown in Figure 5e, which superimposes the relative phase between stimulus and response (solid squares) with the relative phase between stimulus and brain signals from two representative SQUID sensors (open squares; see Kelso *et al.*, 1992 for the full data set). The dotted vertical lines indicates points where the stimulus frequency changed in the experiment. The horizontal lines represent a phase difference of π rad. As to be expected, the SQUID data are somewhat noisier than the behavioural data. Nevertheless, a transition in both brain and behaviour is clearly evident in the relative phase, which typically drifts upward and fluctuates before switching, a definite sign of approaching instability. Critical slowing down is indicated by the fact that both brain and behaviour are perturbed more by the *same* magnitude of perturbation (a step change of 0.25 Hz) as the critical point approaches. It takes longer and longer to return to the pre-perturbation relative phase value as the transition draws near. Pattern formation and switching, in other words, take the form of a dynamical instability. Notably, the coherence of both brain and behavioural signals is captured by the same macroscopic order parameter, relative phase. There is, as it were, an abstract 'order parameter isomorphism' between brain and behavioural events.

In order to characterize the entire spatial array of 37 sensors as it evolves in time a decomposition was performed using the Karhunen–Loève (KL) method (Friedrich, Fuchs & Haken, 1991; Fuchs, Kelso & Haken, 1992). This procedure is also known as principal component analysis or singular value decomposition. The spatiotemporal signal $H(\mathbf{x},t)$ may be decomposed

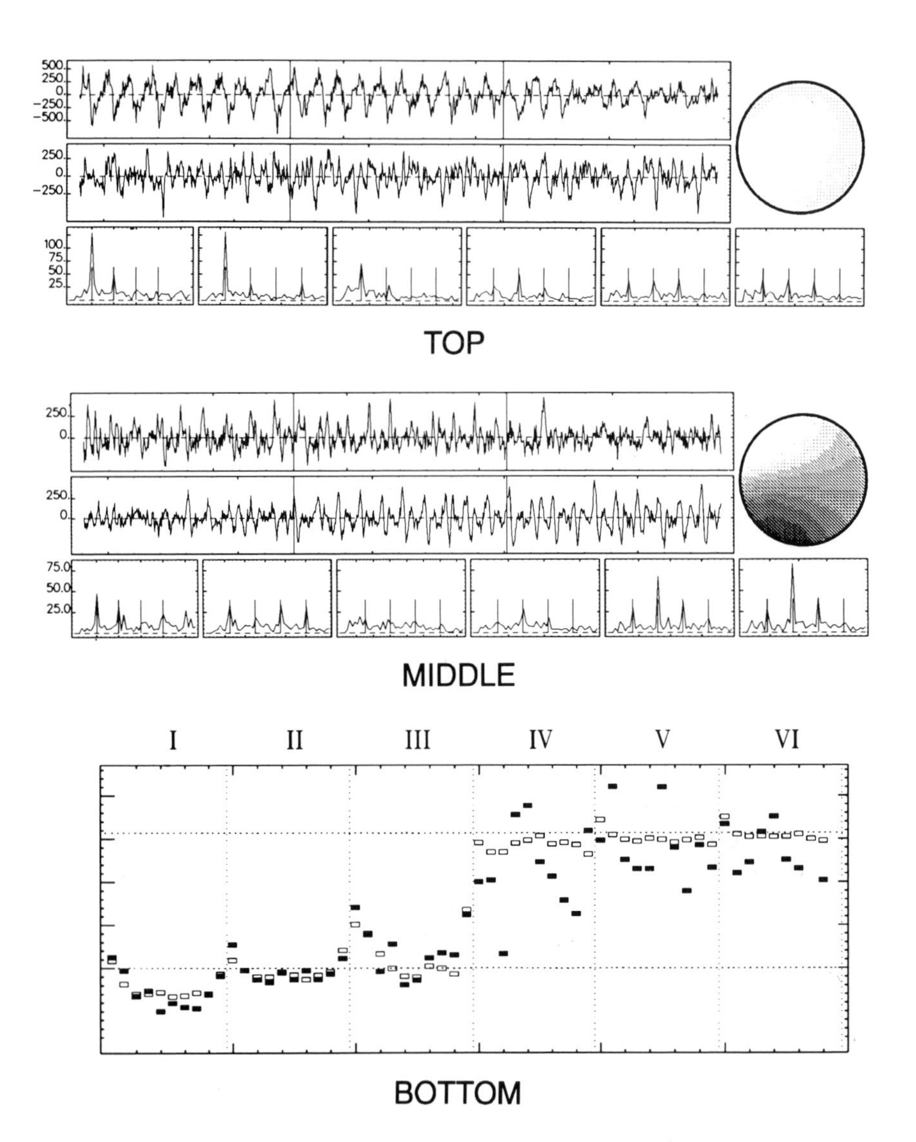

Figure 6. Dynamics of the first two spatial (KL) modes which capture about 75% of the variance in the signal array. *Top*: (right) Dominant KL mode. Amplitudes and power spectra on frequency plateaus I–VI. *Middle*: (right) Second KL mode and corresponding amplitudes and power spectra. *Bottom*: Relative phase of the behaviour (open squares) and the amplitude of the top mode (solid rectangles) with respect to the stimulus. Note the qualitative changes in all three displays around the beginning of plateau IV (see text for details).

into spatial, time-independent modes $\phi_i(\mathbf{x})$ and their corresponding amplitudes $\xi_i(t)$:

$$H(\mathbf{x},t) = \sum_{i=1}^{N} \xi_i(t)\phi_i(\mathbf{x}). \tag{6}$$

If the functions $\phi_i(\mathbf{x})$ are chosen properly, a truncation of this expansion at a small N (say $N< 5 \ldots 10$) gives a good approximation of the original dataset. The KL decomposition is optimal in the sense that it minimizes the mean square error for every truncation point. It turns out that only a few modes are needed to account for most of the variance in the brain signals.

Figure 6 (top, middle) shows the spatial form of the functions obtained by the KL expansion and their amplitudes for the two most dominant modes, i.e. the two largest eigenvalues. For the top mode, which covers about 60% of the power in the signals, a strong periodic component is evident over the entire time series However, the spectra show that there is a qualitative change between pre- and post-transition regions. In the pre-transition régime, the dynamics is dominated by the first KL mode oscillating at the stimulus (and response) frequency of the behaviour. At the transition point a switch occurs, the second KL mode exhibiting a large frequency component at twice the stimulus frequency (Figure 6, middle).

As previously mentioned, the (anti-phase) syncopation pattern is not stable beyond a certain critical frequency, and a spontaneous switch to an (in-phase) synchronization pattern is observed. As shown in Figure 6 (bottom) the first KL mode (solid squares) exhibits a clear transition of π at the transition point. Notice that the phase of the brain signal and the sensorimotor behaviour are almost identical in the pre-transition region, whereas after the transition the brain signal becomes more diffuse even as the sensorimotor behaviour becomes more regular. Relaxational behaviour, typical of critical slowing down is once again evident.

In summary, although the brain possesses tremendous heterogeneity of structure and its dynamics, in general, are nonstationary, it is still possible – under well-defined conditions – to demonstrate its pattern-forming character. From an incoherent spontaneous or 'rest' state the brain manifests coherent spatiotemporal patterns immediately when confronted with a meaningful task. Like many of the complex, nonequilibrium systems studied by synergetics, at critical values of a control parameter, the brain undergoes spontaneous changes in spatiotemporal patterns, measured e.g. in terms of relative phases, spectral properties of spatial modes etc. Remarkably, these quantities exhibit predicted signatures of pattern forming instabilities in self-organizing (synergetic) systems. Current theoretical work is devoted to modelling the dynamics observed here. Empirical studies using full head, 64 sensor arrays have also been undertaken. Sherrington's beautiful image of an enchanted loom where millions of flashing shuttles weave a never abiding but always meaningful pattern is beginning, it seems, to be realized.

CONCLUDING THOUGHTS

Over the years eminent biologists have argued that the methods of doing science on inanimate objects are entirely inadequate for doing science on living things, especially those that have brains and possess intentionality. On the other hand, when eminent physicists come to consider exotic properties of living things like consciousness, they look to ties between physical theories such as quantum mechanics and special relativity for clues. One can only wonder why the physics of cooperative phenomena and self-organization in open, nonequilibrium systems is ignored by both parties. In particular, synergetics and related approaches have shown that over and over again nature uses the same principles to produce 'novel' forms on a macroscopic scale. These are global properties of the system: they are explicitly collective and (usually) quite independent of the material that supports them. Under certain conditions, ordinary matter exhibits extraordinary 'life-like' behaviour, including spontaneous pattern formation, pattern change, and the creation and annihilation of forms. In this paper, only a flavour of the possibilities is given but hopefully these are enough to encourage further exploration of the thesis that living things are *fundamentally* nonequilibrium systems in which new patterns emerge and sustain themselves in a relatively autonomous fashion.

What is it, then, that separates the dead from the alive? Schrödinger proposed ideas such as 'the order from order principle', 'feeding on negative entropy' and the 'aperiodic solid'. Focus on the latter promoted biochemistry and spawned molecular biology, but not much 'new physics'. Yet, it can be argued that open, nonequilibrium systems have much to teach us about the organization of living things – and vice versa. Some of the evidence summarized here shows that living things, including the human brain, tend to 'dwell' in metastable coordinated states poised near instability where they can switch flexibly. They live near criticality where they can anticipate the future and not simply react to the present. All this involves 'new' physics of self-organization, in which, incidentally, no single level is any more, or less, fundamental than any other.

For much of mainstream biology the chief source of biological organization is not its openness but the fact that organisms are controlled by a program. For many geneticists and biologists, the teleonomic character of the organism is due specifically to a *genetic program*. This organisms share with man-made machines and this is what distinguishes them from inanimate nature. According to such views, all that we need to know is that a program exists that is causally responsible for the goal-directedness of living things: how the program originated is quite irrelevant.

The physics of self-organization in open, nonequilibrium systems already provides 'life-like' properties even without a genome. Robert Rosen (1991) has suggested that the free behaviour of open systems is just the kind of thing that Mendelian genes can 'force'. But to describe the gene, even conceptually, as a program sending instructions to cells to organize themselves belittles the complexity of the gene. The more we learn about genetic material, the more the gene itself looks like a self-organized dynamical system. Programs, after all, are written by programmers. Who or what programs the genetic program?

Speculatively, but in true reductionist fashion, the day may come when the very distinction between genotype and phenotype will fade. Even Darwin, and later Lorenz, recognized that behaviour itself emerges from coordinated actions that promote survival of the individual and hence the species. Here and elsewhere it has been shown that certain basic forms of coordination are subject to principles of self-organization. Might, then, the genotype–phenotype relation eventually be construed in terms of shared, self-organized dynamics acting on different time scales? If this is so, we may invoke the slaving principle of synergetics: the slowly varying quantities are the order parameters that enslave the quickly adjusting parts. If the gene pool of a species is considered as slowly varying over the life span of an individual (human, animal or plant), then surely the genes enslave the individual, reminding us of Dawkins's (1976) thesis of the selfish gene. But what will happen if the individual can influence his or her genes? This is at present a quite unorthodox question that implies that Lamarck might once again raise his head. Other order parameters acting on humans are certainly language, culture, science and so on. They, in addition to genes, contribute to the formation of an individual.

REFERENCES

Babloyantz, A. (1986). *Molecule, Dynamics and Life.* New York: Wiley.

Bak, P. (1993). Self-organized criticality and gaia. In *Thinking about Biology*, eds. W. D. Stein & F. J. Varela, pp. 255–268. Reading, MA: Addison Wesley.

Bergé, P., Pomeau, Y. & Vidal, C. (1984). *Order Within Chaos.* Paris: Hermann.

Buchanan, J. J. & Kelso, J. A. S. (1993) Posturally induced transitions in rhythmic multijoint limb movements. *Experimental Brain Research* **94**, 131–142.

Collet, P. & Eckmann, J. P. (1990). *Instabilities and Fronts in Extended Systems.* Princeton, NJ: Princeton University Press.

Collins, J. J. & Stewart, I. N. (1993). Coupled nonlinear oscillators and the symmetries of animal gaits. *Journal of Nonlinear Science* **3**, 349–392.

Crick, F. H. C. (1966). *Of Molecules and Men.* Seattle: University of Washington Press.

Dawkins, R. (1976). *The Selfish Gene.* Oxford: Oxford University Press.

Dyson, F. (1985). *Origins of Life.* Cambridge: Cambridge University Press.

Friedrich, R., Fuchs, A. & Haken, H. (1991). In *Synergetics of Rhythms*, eds. H. Haken & H. P. Köpchen, Berlin: Springer.

Fuchs, A. & Kelso, J. A. S. (1993). Pattern formation in the human brain during qualitative changes in sensorimotor coordination. *World Congress on Neural Networks, 1993* **4**, 476–479.

Fuchs, A., Kelso, J. A. S. & Haken, H. (1992). Phase transitions in the human brain: spatial mode dynamics. *International Journal of Bifurcation and Chaos* **2**(4), 917–939.

Haken, H. (1969). Lecture at Stuttgart University.

Haken, H. (1975). Cooperative phenomena in systems far from thermal equilibrium and in non-physical systems. *Reviews of Modern Physics* **47**, 67–121.

Haken, H. (1977). *Synergetics: An Introduction.* Berlin: Springer.

Haken, H., Kelso, J. A. S. & Bunz, H. (1985). A theoretical model of phase transitions in human hand movements. *Biological Cybernetics* **51**, 347–356.

Ho, M. W. (in press). *The Rainbow and the Worm.* Singapore: World Scientific.

Iberall, A. S. & Soodak, H. (1987). A physics for complex systems. In *Self-organizing Systems: The Emergence of Order*, ed. F. E. Yates. New York and London: Plenum.

Jeka, J. J., Kelso, J. A. S. & Kiemel, T. (1993). Pattern switching in human multilimb coordination dynamics. *Bulletin of Mathematical Biology* **55** (4), 829–845.

Jirsa, V. K., Friedrich, R., Haken, H. & Kelso, J. A. S. (1994). A theoretical model of phase transitions in the human brain. *Biological Cybernetics* **71**, 27–35.

Kauffman, S. A. (1993). *Origins of Order: Self-organization and Selection in Evolution.* Oxford: Oxford University Press.

Kelso, J. A. S. (1981). On the oscillatory basis of movement. *Bulletin of the Psychonomic Society* **18**, 63.

Kelso, J. A. S. (1984). Phase transitions and critical behavior in human bimanual coordination. *American Journal of Physiology: Regulatory, Integrative and Comparative Physiology* **15**, R1000–R1004.

Kelso, J. A. S. (1988). Introductory remarks: Dynamic patterns. In *Dynamic Patterns in Complex Systems*, eds. J. A. S. Kelso, A. J. Mandell & M. F. Shlesinger, pp. 1–5. Singapore: World Scientific.

Kelso, J. A. S. (1990). Phase transitions: foundations of behavior. In *Synergetics of Cognition*, ed. H. Haken, pp. 249–268. Berlin: Springer.
Kelso, J. A. S. (1994). Elementary coordination dynamics. In *Interlimb Coordination: Neural, Dynamical and Cognitive Constants*, eds. S. Swinnen, H. Heuer, J. Massion & P. Casaer. New York: Academic Press.
Kelso, J. A. S., Bressler, S. L., Buchanan, S., DeGuzman, G. C., Ding, M., Fuchs, A. & Holroyd, T. (1991). Cooperative and critical phenomena in the human brain revealed by multiple SQUIDS. In *Measuring Chaos in the Human Brain*, eds. D. Duke & W. Pritchard, pp. 97–112. Singapore: World Scientific.
Kelso, J. A. S., Bressler, S. L., Buchanan, S., DeGuzman, G. C., Ding, M., Fuchs, A. & Holroyd, T. (1992). A phase transition in human brain and behavior. *Physics Letters A* **169**, 134–144.
Kelso, J. A. S., DeGuzman, G. C. & Holroyd, T. (1991). The self-organized phase attractive dynamics of coordination. In *Self-organization, Emerging Properties and Learning, Series B*: Vol 260, ed. A. Babloyantz, pp. 41–62. New York: Plenum.
Kelso, J. A. S., DelColle, J. D. & Schöner, G. (1990). Action-perception as a pattern formation process. In *Attention and Performance XIII*, ed. M. Jeannerod, pp. 139–169. Hillsdale, NJ: Erlbaum.
Kelso, J. A. S., Ding, M. & Schöner, G. (1992). Dynamic pattern formation: a primer. In *Principles of Organization in Organisms*, eds. A. Baskin & J. Mittenthal, pp. 397–439. Redwood City, CA: Addison Wesley.
Kelso, J. A. S. & Jeka, J. J. (1992). Symmetry breaking dynamics of human multilimb coordination. *Journal of Experimental Psychology: Human Perception and Performance*, **18**, 645–668.
Kelso, J. A. S. & Scholz, J. P. (1985). Cooperative phenomena in biological motion. In *Complex Systems: Operational Approaches in Neurobiology, Physical Systems and Computers*, ed. H. Haken, pp. 124–149. Berlin: Springer.
Kelso, J. A. S., Scholz, J. P. & Schöner, G. (1986). Non-equilibrium phase transitions in coordinated biological motion: critical fluctuations. *Physics Letters* **A118**, 279–284.
Kelso, J. A. S., Scholz, J. P. & Schöner, G. (1988). Dynamics governs switching among patterns of coordination in biological movement. *Physics Letters* **A134**(1), 8–12.
Kuramoto, Y. (1984). *Chemical Oscillations, Waves, and Turbulence*. Berlin: Springer.
Maddox, J. (1993). The dark side of molecular biology. *Nature* **363**, 13.
Mayr, E. (1988). *Toward a New Philosophy of Biology*. Cambridge, MA: Harvard University Press.
Moore, W. (1989). *Schrödinger, Life and Thought*. Cambridge: Cambridge University Press.
Nicolis, G. & Prigogine, I. (1989). *Exploring Complexity: An Introduction*. San Francisco: Freeman.
Pattee, H. H. (1976). Physical theories of biological coordination. In *Topics in the Philosophy of Biology*, Vol. 27, eds. M. Grene & E. Mendelsohn, pp. 153–173. Boston: Reidel.
Rosen, R. (1991). *Life Itself.* New York: Columbia University Press.

Schmidt, R. C., Carello, C. & Turvey, M. T. (1990). Phase transitions and critical fluctuations in the visual coordination of rhythmic movements between people. *Journal of Experimental Psychology: Human Perception and Performance* **16**(2), 227–247.

Schmidt, R. C., Shaw, B. K. & Turvey, M. T. (1993). Coupling dynamics in interlimb coordination. *Journal of Experimental Psychology: Human Perception and Performance* **19**, 397–415.

Scholz, J. P., Kelso, J. A. S. & Schöner, G. (1987). Non-equilibrium phase transitions in coordinated biological motion: critical slowing down and switching time. *Physics Letters* **A123**, 390–394.

Schöner, G., Haken, H. & Kelso, J. A. S. (1986). A stochastic theory of phase transitions in human hand movement. *Biological Cybernetics* **53**, 442–452.

Schöner, G., Jiang, W. Y. & Kelso, J. A. S. (1990). A synergetic theory of quadrupedal gaits and gait transitions. *Journal of Theoretical Biology* **142**(3), 359–393.

Schöner, G. & Kelso, J. A. S. (1988a). Dynamic pattern generation in behavioral and neural systems. *Science* **239**, 1513–1520.

Schöner, G. & Kelso, J. A. S. (1988b). A synergetic theory of environmentally-specified and learned patterns of movement coordination. II. Component oscillator dynamics. *Biological Cybernetics* **58**, 81–89.

Schrödinger, E. (1944). *What is Life?* Cambridge: Cambridge University Press.

Shik, M. L., Severin, F. V. & Orlovskii, G. N. (1966). Control of walking and running by means of electrical stimulation. *Biophysics* **11**, 1011.

Swinnen, S., Heuer, H., Massion, J. & Casaer, P. (eds.) (1994). *Interlimb Coordination: Neural, Dynamical and Cognitive Constants*. New York: Academic Press.

von Holst, E. (1939). Relative coordination as a phenomenon and as a method of analysis of central nervous function. In *The Collected Papers of Erich von Holst*, ed. R. Martin, pp. 33–135 (1973). Coral Gables, FL: University of Miami.

Wallenstein, G. V., Bressler, S. L., Fuchs, A. & Kelso, J. A. S. (1993). Spatiotemporal dynamics of phase transitions in the human brain. In *Society for Neuroscience Abstracts*, Vol. 19, p. 1606. Washington DC: Society for Neuroscience.

Wimmers, R. H., Beek, P. J. & van Wieringen, P. C. W. (1992). Phase transitions in rhythmic tracking movements: A case of unilateral coupling. *Human Movement Science* **11**, 217–226.

Wolpert, L. (1991). *The Triumph of the Embryo*. Oxford: Oxford University Press.

Wunderlin, A. (1987). On the slaving principle. *Springer Proceedings in Physics* **19**, 140–147. Berlin: Springer.

Zanone, P. G. & Kelso, J. A. S. (1992). Evolution of behavioural attractors with learning: nonequilibrium phase transitions. *Journal of Experimental Psychology: Human Perception and Performance* **18**/2, 403–421.

12

Order from disorder: the thermodynamics of complexity in biology

ERIC D. SCHNEIDER[1] and JAMES J. KAY[2]

[1]*Hawkwood Institute, Livingston, Montana*

[2]*University of Waterloo, Waterloo, Ontario*

INTRODUCTION

In the middle of the 19th century, two major scientific theories emerged about the evolution of natural systems over time. Thermodynamics, as refined by Boltzmann, viewed nature as decaying towards a certain death of random disorder in accordance with the second law of thermodynamics. This equilibrium seeking, pessimistic view of the evolution of natural systems is contrasted with the paradigm associated with Darwin, of increasing complexity, specialization, and organization of biological systems through time. The phenomenology of many natural systems shows that much of the world is inhabited by nonequilibrium coherent structures, such as convection cells, autocatalytic chemical reactions and life itself. Living systems exhibit a march away from disorder and equilibrium, into highly organized structures that exist some distance from equilibrium.

This dilemma motivated Erwin Schrödinger, and in his seminal book *What is Life?* (Schrödinger, 1944), he attempted to draw together the fundamental processes of biology and the sciences of physics and chemistry. He noted that life was comprised of two fundamental processes; one '*order from order*' and the other '*order from disorder*'. He observed that the gene generated order from order in a species, that is, the progeny inherited the traits of the parent. Over a decade later Watson and Crick (1953) provided biology with a research agenda that has led to some of the most important findings of the last fifty years.

However, Schrödinger's equally important but less understood observation was his order from disorder premise. This was an effort to link biology with the fundamental theorems of thermodynamics (Schneider, 1987). He noted that living systems seem to defy the second law of thermodynamics, which insists that, within closed systems, the entropy of a system should be maximized. Living systems, however, are the antithesis of such disorder. They display marvellous levels of order created from disorder. For instance, plants are highly ordered structures which are synthesized from disordered atoms and molecules found in atmospheric gases and soils.

Schrödinger solved this dilemma by turning to nonequilibrium thermodynamics. He recognized that living systems exist in a world of energy and material fluxes. An organism stays alive in its highly organized state by taking high quality energy from outside itself and processing it to produce, within itself, a more organized state. Life is a far from equilibrium system that maintains its local level of organization at the expense of the larger global entropy budget. He proposed that the study of living systems from a nonequilibrium perspective would reconcile biological self-organization and thermodynamics. Furthermore he expected that such a study would yield new principles of physics.

This paper examines the order from disorder research programme proposed by Schrödinger and expands on his thermodynamic view of life. We explain that the second law of thermodynamics is not an impediment to the understanding of life but rather is necessary for a complete description of living processes. We expand thermodynamics into the causality of the living process and show that the second law underlines processes of self-organization and determines the direction of many of the processes observed in the development of living systems.

THERMODYNAMIC PRELIMINARIES

Thermodynamics has been shown to apply to all work and energy systems including the classic temperature–volume–pressure systems, chemical kinetic systems, electromagnetic and quantum systems. Thermodynamics can be viewed as addressing the behaviour of systems in three different situations: (1) equilibrium (classical thermodynamics), e.g. the actions of large numbers of molecules in a closed system; (2) systems that are some distance from equilibrium, and will return to equilibrium, e.g. molecules in two flasks connected with a closed stopcock; one flask holds more molecules than the other and upon the stopcock being opened the system will come to its equilibrium state of an equal number of molecules in each flask; and (3) systems that have been moved away from equilibrium and are constrained by gradients to be at some distance from the equilibrium state, e.g. two connected flasks with a pressure gradient holding more molecules in one flask than the other.

Exergy is a central concept in our discussion of order from disorder. Energy varies in its quality or capacity to do useful work. During any chemical or physical process the quality or capacity of energy to perform work is irretrievably lost. Exergy is a measure of the maximum capacity of an energy system to perform useful work as it proceeds to equilibrium with its surroundings (Brzustowski & Golem, 1978; Ahern, 1980).

The first law of thermodynamics arose from efforts to understand the relation between heat and work. The first law says that energy cannot be created or destroyed and that the total energy within a closed or isolated system remains unchanged. However, the quality of the energy in the system (i.e. the exergy content) may change. The second law of thermodynamics requires that if there are any processes underway in the system, the quality of the energy (the exergy) in that system will degrade. The second law can also be stated in terms of the quantitative measure of irreversibility, entropy, the change in which is greater than zero for any real process. The second law can also be stated as: any real process can only proceed in a direction which results in an entropy increase.

In 1908 thermodynamics was moved a step forward by the work of Carathéodory (Kestin, 1976) when he developed a proof that showed that the law of 'entropy increase' is not the general statement of the second law. The more encompassing statement of the second law of thermodynamics is that 'In the neighbourhood of any given state of any closed system, there exist states which are inaccessible from it, along any adiabatic path reversible or irreversible.' Unlike earlier definitions this does not depend on the nature of the system, nor on concepts of entropy or temperature.

More recently Hatsopoulos and Keenan (1965) and Kestin (1968) have subsumed the zeroth, first, and second laws into a Unified Principle of Thermodynamics: 'When one isolated system performs a process after the removal of a series of internal constraints, it will reach a unique state of equilibrium: this state of equilibrium is independent of the order in which the constraints are removed.' This describes the behaviour of the second class of systems, which are some distance from equilibrium but are not constrained to be in a nonequilibrium state. The importance of this statement is that it dictates a direction and an end state for all real processes. This statement tells us that a system will come to the equilibrium that constraints permit.

DISSIPATIVE SYSTEMS

These principles outlined above hold for closed isolated systems. However a more interesting class of phenomena belongs to the third class of systems that are open to energy and/or material flows and reside at quasi-stable states some distance from equilibrium (Nicolis & Prigogine, 1977, 1989). Nonliving organized systems (like convection cells, tornadoes, and lasers) and living systems (from cells to ecosystems) are dependent on outside energy fluxes to maintain their organization and dissipate energy gradients to carry out these self-organizing processes. This organization is maintained at the cost of increasing the entropy of the larger 'global' system in which the structure is imbedded. In these dissipative systems, the total entropy change in a system is the sum of the internal production of entropy in the system (which is always greater than or equal to zero), and the entropy exchange with the environment which may be positive, negative, or zero. For the system to maintain itself in a nonequilibrium steady state the entropy exchange must be negative, and equal to the entropy produced by internal processes, such as metabolism.

Dissipative structures which are stable over a finite range of conditions are best represented by autocatalytic positive feedback cycles. Convection cells, hurricanes, autocatalytic chemical reactions, and living systems are all examples of far-from-equilibrium dissipative structures which exhibit coherent behaviour.

The transition in a heated fluid between conduction and the emergence of convection (Bénard cells) is a striking example of emergent coherent organization in response to an external energy input (Chandrasekhar, 1961). In the Bénard cell experiments, the lower surface of a fluid is heated and the upper surface is kept at a cooler temperature. The initial heat flow through the system is by molecule to molecule interaction. When the heat flux reaches a critical value the system becomes unstable and the molecular action of the fluid becomes coherent and convective overturning emerges, resulting in highly structured coherent hexagonal to spiral surface patterns (Bénard cells). These structures increase the rate of heat transfer and gradient destruction in the system. This transition between non-coherent to coherent structure is the system's response to attempts to move it away from equilibrium (Schneider & Kay, 1994). This transition between non-coherent, molecule to molecule heat transfer to coherent structure results in excess of 10^{22} molecules acting in an highly organized manner. This seemingly improbable occurrence is the direct result of the applied temperature gradient and the dynamics of the system at hand, and is the system's response to attempts to move it away from equilibrium.

To deal with this class of nonequilibrium systems we propose a corollary to Kestin's Unified Principle of Thermodynamics. His proof shows that a system's equilibrium state is stable in the Lyapunov sense. Implicit in this conclusion is that a system will resist being removed from the equilibrium state. The degree to which a system has been moved from equilibrium is measured by the gradients imposed on the system.

As systems are moved away from equilibrium, they will utilize all avenues available to counter the applied gradients. As the applied gradients increase, so does the system's ability to oppose further movement from equilibrium.

We shall refer to this as the 'restated second law' and the pre-Carathéodory statements as the classical second law. In chemical systems, Le Chatelier's principle is an example of the restated second law.

Thermodynamic systems exhibiting temperature, pressure, and chemical equilibrium resist movement away from these equilibrium states. When moved away from their equilibrium state they shift their state in a way which opposes the applied gradients and attempt to move the system back towards its equilibrium attractor. The stronger the applied gradient, the greater the effect of the equilibrium attractor on the system. The more a system is moved from equilibrium, the more sophisticated are its mechanisms for resisting being moved from equilibrium. If dynamic and/or kinetic conditions permit, self-organization processes will arise that abet gradient dissipation. This behaviour is not sensible from a classical perspective, but is expected given the restated second law. No longer is the emergence of coherent self-organizing structures a surprise, but rather it is an expected response of a system as it attempts to resist and dissipate externally applied gradients which would move the system away from equilibrium. Hence we have *order emerging from disorder* in the formation of dissipative structures.

So far our discussion has focused on simple physical systems and how thermodynamic gradients drive self-organization. Chemical gradients also result in dissipative autocatalytic reactions, examples of which are found in simple inorganic chemical systems, in protein synthesis reactions, and in phosphorylation, polymerization and hydrolysis autocatalytic reactions. Autocatalytic reaction systems are a form of positive feedback where the activity of the system or reaction augments itself in the form of self-reinforcing reactions. Autocatalysis stimulates the aggregate activity of the whole cycle. Such self-reinforcing catalytic activity is self-organizing and is an important way of increasing the dissipative capacity of the system.

The notion of dissipative systems as gradient dissipators holds for non-equilibrium physical and chemical systems and describes the processes of emergence and development of complex systems. Not only are the processes of these dissipative systems consistent with the restated second law, but it should be expected that conditions permitting, such systems will emerge if there are gradients present. Schrödinger's notion of order from disorder is about the emergence of these dissipative systems, a phenomena which is generally observed in these class 3 thermodynamic systems.

LIVING SYSTEMS AS GRADIENT DISSIPATORS

Boltzmann recognized the apparent contradiction between the heat death of the universe, and the existence of life in which systems grow, complexify, and evolve. He realized that the sun's energy gradient drives the living process and suggested a Darwinian-like competition for entropy in living systems:

> The general struggle for existence of animate beings is therefore not a struggle for raw materials – these, for organisms, are air, water and soil, all abundantly available – nor for energy which exists in plenty in any body in the form of heat (albeit unfortunately not transformable), but a struggle for entropy, which becomes available through the transition of energy from the hot sun to the cold earth. (Boltzmann, 1886.)

Boltzmann's ideas were further explored by Schrödinger, who noted that some systems, like life, seem to defy the classical second law of thermodynamics (Schrödinger, 1944). However, he recognized that living systems are open and not the adiabatic closed boxes of classical thermodynamics. An organism stays alive in its highly organized state by importing high quality energy from outside itself and degrading it to support the organizational structure of the system. Or as Schrödinger said, the only way a living system stays alive, away from maximum entropy or death, is

> by continually drawing from its environment negative entropy ... Thus the device by which an organism maintains itself stationary at a fairly high level of orderliness (= fairly low level of entropy) really consists in continually sucking orderliness from its environment ... plants ... of course, have their most powerful supply of 'negative entropy' in the sunlight. (Schrödinger, 1944.)

Life can be viewed as a far-from-equilibrium dissipative structure that maintains its local level of organization, at the expense of producing entropy in the environment.

If we view the earth as an open thermodynamic system with a large gradient impressed on it by the sun, the restated second law suggests that the system will reduce this gradient by using all physical and chemical processes available to it. We suggest that life exists on earth as another means of dissipating the solar induced gradient and, as such, is a manifestation of the restated second law. Living systems are far-from-equilibrium dissipative systems and have great potential for reducing radiation gradients on earth (Kay, 1984; Ulanowicz & Hannon, 1987).

The origin of life is the development of another route for the dissipation of induced energy gradients. Life ensures that these dissipative pathways continue and has evolved strategies to maintain these dissipative structures in the face of a fluctuating physical environment. We suggest that living systems are dynamic dissipative systems with encoded memories, the genes, that allow dissipative processes to continue.

We have argued that life is a response to the thermodynamic imperative of dissipating gradients (Kay, 1984; Schneider, 1988). Biological growth occurs when the system adds more of the same types of pathways for degrading imposed gradients. Biological development occurs when new types of pathways for degrading imposed gradients emerge in the system. This principle provides a criterion for evaluating growth and development in living systems.

Plant growth is an attempt to capture solar energy and dissipate usable gradients. Plants of many species arrange themselves into assemblies to increase leaf area so as to optimize energy capture and degradation. The gross energy budgets of terrestrial plants show that the vast majority of their energy use is for evapotranspiration, with 200–500 grams of water transpired per gram of fixed photosynthetic material. This mechanism is a very effective energy degrading process with 2500 joules used per gram of water transpired

(Gates, 1962). Evapotranspiration is the major dissipative pathway in terrestrial ecosystems.

The large scale biogeographical distribution of species richness is strongly correlated with potential annual evapotranspiration (Currie, 1991). These strong relationships between species richness and available exergy suggest a causal link between biodiversity and dissipative processes. The more exergy available to be partitioned among species the more pathways there are available for energy degradation. Trophic levels and food chains are based upon photosynthetic fixed material and further dissipate these gradients by making more highly ordered structures. Thus we would expect more species diversity to occur where there is more available exergy. Species diversity and trophic levels are vastly greater at the equator, where 5/6 of the earth's solar radiation occurs, and there is more of a gradient to reduce.

A THERMODYNAMIC ANALYSIS OF ECOSYSTEMS

Ecosystems are the biotic, physical, and chemical components of nature acting together as nonequilibrium dissipative processes. Ecosystem development should increase energy degradation if it follows from the restated second law. This hypothesis can be tested by observing the energetics of ecosystem development during the successional process or as they are stressed.

As ecosystems develop or mature they should increase their total dissipation, and should develop more complex structures with greater diversity and more hierarchical levels to assist in energy degradation (Schneider, 1988; Kay & Schneider, 1992). Successful species are those that funnel energy into their own production and reproduction and contribute to autocatalytic processes, thereby increasing the total dissipation of the ecosystem.

Lotka (1922) and Odum and Pinkerton (1955) have suggested that those biological systems that survive are those that develop the most power inflow and use it to best meet their needs for survival. A better description of these 'power laws' may be that biological systems develop so as to increase their energy degradation rate, and that biological growth, ecosystem development, and evolution represent the development of new dissipative pathways. In other words ecosystems develop in a way which increases the amount of exergy that they capture and ultilize. As a consequence, as ecosystems develop, the exergy of the outgoing energy decreases. It is in this sense that ecosystems develop the most power, that is, they make the most effective use of the exergy in the incoming energy while at the same time increasing the amount of energy they capture.

This theory suggests that disorganizing stresses will cause ecosystems to retreat to configurations with lower energy degradation potential. Stressed ecosystems often appear similar to earlier successional stage ecosystems and reside closer to thermodynamic equilibrium.

Ecologists have developed analytical methods that allow analysis of material and energy flows through ecosystems (Kay, Graham & Ulanowicz, 1989). With these methods it is possible to detail the energy flow and how the energy is partitioned in the ecosystem. We have recently analysed a data set for carbon and energy flows in two aquatic tidal marsh ecosystems adjacent to a large power generating facility on the Crystal River in Florida (Ulanowicz, 1986). The ecosystems in question were a 'stressed' and a 'control' marsh. The 'stressed' ecosystem is exposed to hot water effluent from the nuclear power station. The 'control' ecosystem is not exposed to the effluent but is otherwise exposed to the same environmental conditions. In absolute terms all the flows dropped in the stressed ecosystem. The implication is that the stress has resulted in the ecosystem shrinking in size, in terms of biomass, its consumption of resources, material and energy cycling, and its ability to degrade and dissipate incoming energy.

Overall the impact of the effluent from the power station heating water has been to decrease the size of the 'stressed' ecosystem and its consumption of resources while impacting on its ability to retain the resources it has captured. This analysis suggests that the function and structure of ecosystems follow the development path predicted by the behaviour of nonequilibrium thermodynamic structures and the application of these behaviours to ecosystem development patterns.

The energetics of terrestrial ecosystems provides another test of the thesis that ecosystems will develop so as to degrade energy more effectively. More developed dissipative structures should degrade more energy. Thus we expect a more mature ecosystem to degrade the exergy content of the energy it captures more completely than a less developed ecosystem. The exergy drop across an ecosystem is related to the difference in black body temperature between the captured solar energy and the energy reradiated by the ecosystem. If a group of ecosystems are bathed by the same amount of incoming energy, we would expect that the most mature ecosystem would reradiate its energy at the lowest exergy level; that is, the ecosystem would have the coldest black body temperature.

Luvall and Holbo (1989, 1991) have measured surface temperatures of various ecosystems using a thermal infrared multispectral scanner (TIMS). Their data show one unmistakable trend, that when other variables are constant the more developed the ecosystem, the colder its surface temperature and the more degraded its reradiated energy.

TIMS data from a coniferous forest in western Oregon showed that ecosystem surface temperature varies with ecosystem maturity and type. The highest temperatures were found at a clear-cut and over a rock quarry. The coldest site, 299K, some 26K colder than the clear-cut, was a 400-year-old mature Douglas Fir forest with a three tiered plant canopy. A quarry degraded 62% of the net incoming radiation while the 400-year-old forest degraded 90%. Intermediate-aged sites fell between these extremes, increasing energy degradation with more mature or less perturbed ecosystems. These unique data sets show that ecosystems develop structure and function that degrades imposed energy gradients more effectively (Schneider & Kay, 1994).

Our study of the energetics of ecosystems treats them as open systems with high quality energy pumped into them. An open system with high quality energy pumped into it can be moved away from equilibrium. But nature resists movement away from equilibrium. So ecosystems, as open systems, respond, whenever possible, with the spontaneous emergence of organized behaviour that consumes the high quality energy in building and maintaining the newly emerged structure. This dissipates the ability of the high quality energy to move the system further away from equilibrium. This self-organization process is characterized by abrupt changes that occur as a new set of interactions and activities by components and the whole system emerge. This emergence of organized behaviour, the essence of life, is now understood to be expected by thermodynamics. As more high quality energy is pumped into an ecosystem, more organization emerges to dissipate the energy. Thus we have *order* emerging from *disorder* in the service of causing even more disorder.

ORDER FROM DISORDER AND ORDER FROM ORDER

Complex systems can be classified on a continuum of complexity from ordinary complexity (Prigoginean systems, tornadoes, Bénard cells, autocatalytic reaction systems) to emergent complexity perhaps including human socio-economic systems. Living systems are at the more sophisticated end of the continuum. Living systems must function within the context of the system and environment they are part of. If a living system does not respect the circumstances of the supersystem it is part of, it will be selected against. The supersystem imposes a set of constraints on the behaviour of the system and living systems which are evolutionarily successful have learned to live within them. When a new living system is generated after the demise of an earlier one, it would make the self-organization process more efficient if it were constrained to variations which have a high probability of success. Genes play this role in constraining the self-organization process to those options which have a high probability of success. They are a record of successful self-organization. Genes are not the mechanism of development; the mechanism is self-organization. Genes bound and constrain the process of self-organization. At higher hierarchical levels other devices constrain the self-organization process. The ability of an ecosystem to regenerate is a function of the species available for the regeneration process.

Given that living systems go through a constant cycle of birth–development–regeneration–death, preserving information about what works and what does not is crucial for the continuation of life (Kay, 1984). This is the role of the gene and, at a larger scale, biodiversity: to act as information databases of self-organization strategies that work. This is the connection between the order from order and order from disorder themes of Schrödinger. Life emerges because thermodynamics mandates order from disorder whenever sufficient thermodynamic gradients and environmental conditions exist. But if life is to continue, the same rules require that it be able to regenerate, that is, create order from order. Life cannot exist without both processes, *order from disorder to generate life* and *order from order to ensure the continuance of life.*

Life represents a balance between the imperatives of survival and energy degradation. To quote Blum (1968):

> I like to compare evolution to the weaving of a great tapestry. The strong unyielding warp of this tapestry is formed by the essential nature of elementary non-living matter, and the way in which this matter has been brought together in the evolution of our

planet. In building this warp the second law of thermodynamics has played a predominant role. The multi-colored woof which forms the detail of the tapestry I like to think of as having been woven onto the warp principally by mutation and natural selection. While the warp establishes the dimensions and supports the whole, it is the woof that most intrigues the aesthetic sense of the student of organic evolution, showing as it does the beauty and variety of fitness of organisms to their environment. But why should we pay so little attention to the warp, which is after all a basic part of the whole structure? Perhaps the analogy would be more complete if something were introduced that is occasionally seen in textiles, the active participation of the warp in the pattern itself. Only then, I think, does one grasp the full significance of the analogy.

We have tried to show the participation of the warp in producing the tapestry of life. To return to Schrödinger, life is comprised of two processes, order from order, and order from disorder. The work of Watson and Crick and others described the gene, and solved the order from order mystery. This work supports Schrödinger's order from disorder premise and better connects macroscopic biology with physics.

REFERENCES

Ahern, J. E. (1980). *The Exergy Method of Energy Systems Analysis.* New York: Wiley.

Blum, H. G. (1968). *Time's Arrow and Evolution.* Princeton: Princeton University Press.

Boltzmann, L. (1886). The second law of thermodynamics. Reprinted (1974) in *Ludwig Boltzmann, Theoretical Physics and Philosophical Problems*, ed. B. McGuinness. New York: D. Reidel.

Brzustowski, T. A. & Golem, P. J. (1978). Second law analysis of energy processes. Part 1: Exergy – an introduction. *Transactions of the Canadian Society of Mechanical Engineers*, 4(4), 209–218.

Chandrasekhar, S. (1961). *Hydrodynamics and Hydromagnetic Stability.* London: Oxford University Press.

Currie, D. (1991). Energy and large-scale patterns of animal-and-plant species-richness. *American Naturalist* **137**, 27–48.

Gates, D. (1962). *Energy Exchange in the Biosphere.* New York: Harper and Row.

Hatsopoulos, G. & Keenan, J. (1965). *Principles of General Thermodynamics.* New York: Wiley.

Kay, J. J. (1984). Self-Organization in Living Systems. Ph.D. thesis, Systems Design Engineering, University of Waterloo, Ontario.

Kay, J. J., Graham, L. & Ulanowicz, R. E. (1989). A Detailed Guide to Network Analysis. In *Network Analysis in Marine Ecosystems*, eds. F. Wulff, J. G. Field & K. H. Mann, Coastal and Estuarine Studies, Vol. 32, pp. 16–61. New York: Springer-Verlag.

Kay, J. & Schneider, E. (1992). Thermodynamics and measures of ecosystem integrity.

In *Ecological Indicators*, eds. D. McKenzie, D. Hyatt & J. McDonald, pp. 159–181. New York: Elsevier.

Kestin, J. (1968). *A Course in Thermodynamics*. New York: Hemisphere Press.

Kestin, J. (ed.) (1976). *The Second Law of Thermodynamics*. Benchmark Papers on Energy, Vol. 5. Investigations into the foundations of thermodynamics, by C. Carathéodory, pp. 225–256. New York: Dowden, Hutchinson, and Ross.

Lotka, A. (1922). Contribution to the energetics of evolution. *Proceedings of the National Academy of Sciences USA* **8**, 148–154.

Luvall, J. C. & Holbo, H. R. (1989). Measurements of short term thermal responses of coniferous forest canopies using thermal scanner data. *Remote Sensing of the Environment* **27**, 1–10.

Luvall, J. C. & Holbo, H. R. (1991). Thermal remote sensing methods in landscape ecology. In *Quantitative Methods in Landscape Ecology*, eds. M. Turner & R. H. Gardner, Chap. 6. New York: Springer-Verlag.

Nicolis, G. & Prigogine, I. (1977). *Self-Organization in Nonequilibrium Systems*. New York: Wiley.

Nicolis, G. & Prigogine, I. (1989). *Exploring Complexity*. New York: Freeman.

Odum, H. T. & Pinkerton, R. C. (1955). Time's Speed Regulator. *American Scientist* **43**, 321–343.

Schneider, E. D. (1987). Schrödinger shortchanged. *Nature* **328**, 300.

Schneider, E. (1988). Thermodynamics, information, and evolution: new perspectives on physical and biological evolution. In *Entropy, Information, and Evolution: New Perspectives on Physical and Biological Evolution*, eds. B. H. Weber, D. J. Depew & J. D. Smith, pp. 108–138. Boston: MIT Press.

Schneider, E. & Kay J. (1994) Life as a manifestation of the second law of thermodynamics. *Mathematical and Computer Modeling* **19**, nos. 6–8, 25–48.

Schrödinger, E. (1944). *What is Life?* Cambridge: Cambridge University Press.

Ulanowicz, R. E. (1986). *Growth and Development: Ecosystem Phenomenology*. New York: Springer.

Ulanowicz, R. E. & Hannon, B. M. (1987). Life and the production of entropy. *Proceedings of the Royal Society B* **232**, 181–192.

Watson, J. D. & Crick, F. H. C. (1953). Molecular structure of nucleic acids. *Nature* **171**, 4356, 737–738.

13

Reminiscences

RUTH BRAUNIZER
Alpbach, Tirol

I would like to stress that I am not a scientist, so I have accepted the invitation as an elegant gesture of loyalty towards my father to honour his memory. You will therefore excuse me for not referring to my father's work.

At a similar occasion last year in Paris I was asked to contribute biographical notes about my father and had to express my doubts concerning biographies in general. They very often give only their author's views and serve *his* purposes. They rarely do justice to the subjects themselves and tend to typecast them in the eyes of the public. They stick out like monuments until someone suddenly takes pleasure in pointing out their weaknesses and shortcomings, as if these actually had any meaning at all.

In our age voyeurism is very much in vogue and hardly any figure of public life, whether he be of genuine importance or not, can escape it. Anyway, a true narrative of Erwin Schrödinger's life has yet to be written. In order to be true it would have to deal with the facts *only* and forego fiction and any catering to the public taste.

At this point I am grateful for a quotation from Albert Einstein: 'The essential of the being of a man of my type lies precisely in *what* he thinks and *how* he thinks, *not* what he does or suffers'. What Erwin Schrödinger thought and how he thought is for the greatest part common knowledge in the world of physics, and everyone who understands his language can read it, rethink it, interpret it and, if he so wishes, contradict or support it. It is not for me to join in this game. What we all cannot guess is what made him think and think that way. If we could come up with an explanation for it, it would mean that we knew the answer to the fundamental question of life. For me, even to attempt it would be pretentious. What I can do, however,

This chapter is based on a speech delivered at the symposium banquet by Mrs Ruth Braunizer, the daughter of Erwin Schrödinger is go back in time and take a look at the decisive influences he was exposed to in his life, and try to remember what he wanted for himself.

The main influence was the milieu of Vienna between the turn of the century and the end of the 1920s. Not having been a witness myself I could only listen with fascination to the tales of older people when they spoke of those times. De Toqueville once remarked that no one who had not lived before the French Revolution could imagine what life was like then; something similar can be said about the last decades of the Austrian Empire. There was a rapid growth of intellectual brilliance and talent in almost every field; dozens of famous names could be mentioned at this point. Vienna University was a Mecca for so many. There was the Austrian School of Economics, the Vienna School of Medicine, there were the painters, the composers, the architects, the sculptors, the writers and actors.

The still waters of the waning empire provided a breeding ground for almost everything that came to life, and not least for the still widely unknown community of theoretical physicists. There was an excellent school system which stressed the humanities and being inexpensive it offered opportunities to all children, including those of moneyless parents.

The result was a relatively large social group of thoroughly educated people, men as well as women. A member of that generation, whatever their occupation, whether medical doctor, public servant, engineer or sea-captain would, for instance, have been able to enjoy Plato or Seneca in the original, without seeking the help of a dictionary or a commentary. Consequently, they were also masters of their own language. This came to my mind recently, when upon receiving and reading the letter of a young physicist, I did so with growing astonishment and disbelief. It was so full of grammatical and spelling errors that I wondered how he had got even as far as high-school, not to speak of further levels. Yet, he is a most promising scientist, highly thought of by his peers. Presumably, in the days of my father he would not have got to where he is, the system would have rejected him much earlier, or forced him to do his homework.

Obviously in our time you can get ahead without bothering about culture. The incident raises several questions. Are we just looking at the result of specialization? My father dreaded specialization and strove to be a generalist in every way. But this was the mark of his generation. Was it, though, beyond that, not also a very personal thing with him, something essential for his progress? Or is the young physicist who I mentioned just giving proof that a genius will prevail under any circumstances.

Anyway, my father would not have been admitted to *Gymnasium* much less to University without perfect grammar and perfect spelling. His genius would have been relegated to a different level, perhaps that of a cultural freelance. He could maybe have become a famous painter, hardly a famous writer, who knows?

But after stressing the importance of higher education in those times it is only fair to point out that in 1914 nearly all major countries were governed by a set of very well educated and highly cultured people who, in spite of their learning, led mankind into the hitherto greatest catastrophe. Contemplating all this, I must come to the conclusion that, whether his education and fine cultural grooming were relevant or not to his scientific achievement, they were certainly essential to his appearance and to the impression he made as a human being. He was a gentleman of the old times, which made him a very pleasant and lovable man to live with. He made one long for times gone by.

Having said that, we cannot overlook his parents' guidance, which was indeed very influential. His English-bred mother's bilinguality and family-ties were soon to become his as well. His mother loved music and played the violin beautifully. When she died of breast cancer at the age of 54, her son maintained it could have been aggravated by her fervent practice of the instrument. Her death, together with his father's death nearly two years before, left a tragic imprint on Erwin Schrödinger. From that time on he broke off any relationship to music he had had up till then.

His father ran a family business of manufacturing and distributing special cloth, but at heart he was a biologist, or scientist, and, besides, very, very interested in art. In the true French sense he was a dilettante, this being something positive. It describes a person of talent and intellect, keen in pursuing knowledge in fields of interest outside one's profession. Father Schrödinger also possessed a large library, which his son used at random practically from the day he was able to read. One of the few true regrets I heard my father voice later on was the loss of that library, which, in a moment of carelessness, he had decided to sell after his father's death.

People who are considered outstanding and who have consequently become famous run the risk of becoming legendary. Sometimes such legends are found out to be false by some eager historian. Generations of school-children in the German-speaking world learnt that Goethe's last words were 'More Light'. Now we are being told that, quite differently, he addressed a young lady, saying 'little one, hold my hand once more'. Legends, when destroyed, are often replaced by other legends. Even in the close circle of a family a legendary image of a deceased person tends to develop. It is difficult to stay clear of such transformations of memory.

Here it may help to recall bits and pieces of conversations and other exchanges that give a clue of what the person thought of himself, or what image he would paint of himself, if asked. I remember rather vividly one such conversation, which took place about two years before my father's death. It was based on somebody else's progress and on what subject he or she should choose for their future studies. There my father quite abruptly and emphatically said: 'Before I ever knew what subject I should choose, I had made up my mind to become a teacher'. This sentence, although chiselled into my memory, is not a legend. It is rather a glimpse of the real Erwin Schrödinger. Not only was he, as I have learnt from many of his students, a very good teacher, having a beautifully clear and simple way of expressing himself, when speaking as well as writing – his multilingual upbringing must have helped – but beyond that, the teaching-profession meant something else in his life. He needed it to do what he did; it was truly instrumental.

I am certain there are a great many people, many millions perhaps, who often have worthwhile and wonderful thoughts and ideas. Splendid theories, maybe marvellous solutions to many problems are enshrined in the minds and skulls of thousands every day. The only trouble is, they are never brought forward, so that thoughts and ideas are lost again. They may not be recognized by the bearer as something special, or he may not be able to announce them. The teaching-profession itself did not make my father have his brain-waves, as teachers have seldom anything more remarkable to say than other people. But his original urge to use this most efficient vehicle for transmitting whatever ideas he might have was probably part of the force that drove him.

When we came to Ireland over fifty years ago we were refugees. The numbers of refugees may hardly have changed since then, but the times have done so quite considerably, and we have reason to believe that we are watching the closing scenes of an unhappy age of violence. My father, having been a refugee himself, would sympathize with all those who have to leave their homeland in order to save their lives. He became a refugee because of his outspoken opposition to the Nazi-regime. If not for that he could have stayed and been one of Hitler's sages who survived the war unharassed and had but little trouble or regrets afterwards.

Unlike millions of poor people who were persecuted for their birth he had a choice. He chose to leave. And unlike so many others we were privileged. We did not have to beg to be accepted in a foreign country or fear we would be turned away. We were invited and offered generous hospitality. For this we are forever grateful to Ireland, to her people and to Éamonn de Valera, one of my father's greatest friends.

I have said this time and again in the past and I am very happy to repeat it today, over half a century after we came to Dublin on this happy occasion in this Fair City.

Index

Note: page numbers in *italics* refer to figures